Walking with Giants

Europe's Massive Earthmovers

Walking with
Giants

Europe's Massive Earthmovers

Steven Vale

Old Pond
PUBLISHING LTD

Published by

Old Pond Publishing Ltd
Dencora Business Centre, 36 White House Road
Ipswich IP1 5LT United Kingdom

www.oldpond.com

Pictures printed with kind permission of Caterpillar Inc, Bucyrus,
Komatsu Mining Germany, Liebherr, Port of Rotterdam Authority/PUMA,
Aeroview by Dick Sellenraad, and Simon Ramella and Florian Rossero

Cover design and book layout by Liz Whatling
Printed and bound in China

Contents

Dedication

Today's modern hydraulic excavators are a world apart from the monstrous walking draglines that once dominated the skyline of Britain's coalfields during the heydays of the 1960s, 1970s and 1980s.

This book is dedicated to the memory of all those who worked with this impressive fleet; in particular, to the Friends of St Aidan's BE1150 Walking Dragline, a small group of enthusiastic volunteers who worked tirelessly to save and preserve one of the last survivors – a sixty-two-year-old Bucyrus-Erie 1150-B, called Oddball.

The Friends of St Aidan's BE1150 Walking Dragline – headed by Paul Thompson, John Hopkinson, Dave Weedon and Ivor Brown – have to be credited with doing all the hard work to save the last great walking dragline, but two strokes of good fortune also played a part. The first stroke of luck was the donation of the machine to the St Aidan's Trust by RJB Mining, which could have opted for a lucrative scrap – a value ranging from £75,000 to £150,000 – instead.

The second stroke of good luck was a £15,000 donation from the PRISM fund of the Science Museum, which paved the way for another donation of £100,000 of National Lottery money to secure the future of this truly amazing earthmover – believed to be the last of its kind in Western Europe.

This cash injection allowed the volunteers to move the immense machine to its final resting place – an approximate cost of £70,000 – and also left enough to restore Oddball to its original Coal Board livery.

Although the enthusiasts managed to save the carcass, it is saddening to know that Oddball will never raise its boom again, because many of the copper components were plundered when it was first parked.

Also disheartening is the uncertain future of the other two active but old draglines in Europe – a Marion 7400 working in Sweden and another working in France.

Still, on a brighter note, it is encouraging that Oddball is preserved at its final resting place in Swillington, near Leeds, for generations to come – a little piece of history which has already been visited by several thousand people.

Acknowledgements

My walk with giants started in 2004 with the launch of a new British magazine called *Earthmovers*. At the time I was living – and still am – in the Netherlands and working as a European correspondent for a sister title called *Classic Tractor*.

My first assignment for *Earthmovers* was to cover the Bauma show, which is held every three years in Germany. It is one of the biggest machinery events in the world. With little previous knowledge of construction machinery, I found it to be a tough assignment.

Bauma is not for the faint-hearted. The site sprawls over 50 hectares and attracts around half a million visitors. I was asked if I would attend only a few days before it started and I quickly discovered that all hotels within a 50 mile radius of the city were fully booked months – possibly years – in advance, with the nearest available accommodation being almost 100 miles from Munich.

Overwhelmed and a little out of my depth, I managed to take one thing home from my first taste of the earthmoving world – the big machines. Liebherr's massive T 282 dump truck and R 994B hydraulic excavator ignited a spark inside me and I became determined to discover more about these, and other, mega earthmovers.

Over the past seven years, I have been extremely fortunate to be able to spend valuable time at many of the biggest surface mines in the world. This was thanks not just to *Earthmovers* for whom I wrote many an article but also Old Pond Publishing's film crew who I helped to produce the Massive Machines DVD series.

Travelling from Southern Spain to north of the Arctic Circle, my journey has allowed me to document and witness an elusive group of some of the largest man-made machines on the planet. Many of them operate in quite remote locations, while others are, surprisingly, not that far from densely populated areas. I do not pretend to cover every single machine in Europe. Neither is this book a historical reference to every large earthmover that has ever graced the European scene. Instead, it is a mix of some older highlights, with all the biggest, most important and newest machines in Europe today.

I am not just a mining fan but also a traveller – a sort of mining tourist. In fact, I had originally planned to call this book 'Mining Tourist' but while I toured Europe in my effort to document the largest earthmovers I realised that I was more than just a tourist. I was walking with these machines, and so that seemed a more apt title.

My walk has brought me close to the unbelievably large bucket wheel excavators in Germany; close to the big powerful Liebherr excavators in the stunning environment of the French Pyrenees Mountains; and close to the monstrous machines in the frozen forests north of the Arctic Circle at Sweden's Aitik copper mine.

I was fortunate to get a close look at all these awesome machines – and many others – but I could not have done so without a great deal of help and assistance from several hundred individuals. Without their help, there would have been no book in the first place.

It is impossible to mention everyone, but there are numerous people I would like to thank for their kind help in the compilation of this book. I am particularly grateful to the following: Jan Arild Aaserud, Peet Adams, Thomas Åhman, Randy Aneloski, Penin Alvaro, Miguel Angel, Alberto Ayerbe, Kevin Banks, Peter Beatie, Bernard Beck, Dennis van Beek, Darren Bennett, Fred van Bijnen, Christophe Binsse, Konrad Böcker, Carolien Bögemann, Lau Boisen, Milco Bolognesi, Swann Blaise, Ivor Brown, Peter Burck, Michelle Cairns, Francisco de J. Canelo, Coral Castañon, Chris Chamberlain, Constant Charier, Julien Conté, Finn Corwin, Brian Coulson, Colin Cox, John Cunningham, Clifford Daly, Jason Paul Davies, Hywel Davies, René Deubelbeiss, Jackie Dover, Mark Dowdall, Ian Dryburgh, Phil Dukes, Malcolm Edwards, Camilla Enback, Mikael Engelmark, Joakim Eriksson, Wayne Evans, Paul Evans, Kenneth Ewart, Pilar Fernandez, Dal Castello Filippo, Andrew Foster, Philippe Fritz, Sophie Fromont, Mick Fulham, Luis Garcia, Antonio Girei, Daniel Grab, Marc Gulikers, Priscilla van der Haar, Peter Haddock, Anna Hagström, Eric Halleux, Lynsey Hanney, Glenn Hannu, Börje Hansson, Ken Hannah, Richard Halderthay, Mark Harrington, Mark Heames, Alan Hiddleston, Evert Hof, Roger Hollinworth, John Hopkinson, Dave Hurcombe, Kent Isaksson, Catrin Isaksson, Anna Johansson, Arvid

Jhonnsen, Kylie Jones, Laura Jones, Aarno Juntura, Seppo Kankaanpää, Mia Karlson, Magnus Keskitalo, Tracey Kirkup, Jo Kleijnen, Franziska Knoblich, Brian Lamb, José Lambert, David Lancashire, Mikael Larsson, Niklas Larsson, Philippe Lehman, Ronny Maes, Justo Martin, Gerd Martinsson, David Marston, Gary Meyer, Peter Meyer, Inmaculada Montero, César Mulliez, Jari Nygren, Per Österman, Stuart Oliver, Robert Orr, Vincent van Overbeke, Christelle Paillou, Leif Palo, Álvaro Penin, Katie Perkin, Francesco and Roberto Perlini, Mireille Pijck, Martyn Pitt, Daniel Poll, Sjaak Poppe, James Poyner, Edward Prosser, Juan Ramon, David M. Rea, Winfried (Winni) Rechenberg, Arne Renström, Huw Richards, Juha Riikonen, Juan José Gago Rodríguez, Carl Rowley, Keith Rowley, Geert-Yke Rusticus, Fabrice Santoro, John Sammut, Manuel Santamaria, Tim Schlink, Erland Segerstedt, Felicia Sigglow, Luis Simon, Colin Slater, Derek Smibert, Allen Smith, Klaas Smits, Grethe Spongsveen, Mark Sprouls, Anders Stene, Johanna Storm, Ellie Swinbank, Luis Tahaocas, Steve Ternent, Paul Thompson, Sylvia Töllner, Bas Tolhuizen, Sigurd Trones, Juanjo Valverde, Janier Vasquez, Ann Verheeke, Laurent Vermeulen, Ernesto de Vicente, Gisèle Vinandy, Tonia Vogt, Mark Vreeburg, Gary Ward, Jacqueline Walton, Jean-Marc Wellens, Lisa Llewellyn-Williams and Charlie Zimmerman.

I would like to extend a special thanks to Trevor Meek at UK Coal, Arto Suokas at Talvivaara in Finland, Knut Petter Netland at Titania in Norway, Dirk Tegtmeier at Bucyrus and Erik Zimmermann at Komatsu Mining Germany. I am also grateful to Guido Stefan for providing some truly spectacular images of RWE Power's bucket wheel excavators.

I would also like to thank all the staff at the Aitik copper mine for making me feel extremely welcome during my four-week-long visits over the past seven years. I am particularly grateful to Glenn Nilsson and Sture Holgrem for their support and willingness to help. I am also deeply indebted to my friend and former Aitik employee Torbjörn Kjellsson, for patiently sharing with me his vast knowledge of the Swedish mine.

It would be very remiss of me to forget the enthusiastic team at Victorino Alonso's Spanish coal mines, who went out of their way to help me gather as much information and as many pictures as possible. Special thanks are also due to Roberto López Braña for welcoming me back on five different occasions during the past few years to chart the development of his ageing fleet of Caterpillar excavators and the amazing twenty-eight-strong fleet of the largest Hitachi excavators in Europe.

I have found four books particularly helpful when I was researching the background: *Opencast Coal & Equipment*, Bill Huxley (Roundoak); *Scotland's Black Diamonds*, Guthrie Hutton (Stenlake Publishing); *Sunshine Miners*, Peter N Grimshaw; *Giant Earthmovers*, Keith Haddock (Crestline).

My walk with giants has taken me to fourteen different countries and allowed me to witness some truly spectacular scenery. After several dozen flights and many thousands of kilometres behind the wheel of a hire car, this book is the culmination of seven years of travelling.

I stand in awe of the enthusiastic French duo of Simon Ramella and Florian Rossero who recently completed a similar journey by car visiting many of the same sites and machines covered in this book. They managed to pack it all into a whirlwind five-week and 23,000 km journey! I am extremely grateful to them for sharing their extensive knowledge and some of their wonderful French pictures.

I would like to thank *Earthmovers* editor Graham Black for his continued support, and film director Jonathan Theobald and cameraman Roger Wiltshire for their endless patience while filming the 'Massive Machines' series of DVDs.

Finally, thanks to publisher Roger Smith of Old Pond, not only for his advice and financial backing for the 'Massive Machines' DVDs, but also for giving me the green light to tackle this latest hugely interesting project; and thanks also to Georgina Ferrier and Liz Whatling for pulling it together.

Now that it is finished, I am already planning further jaunts deeper into Eastern and Central Europe. First though, I am looking forward to spending more time with my wife Hilda, and my two children Carmina and Cameron.

Steven Vale
De Rijp, The Netherlands, 2011

Introduction

The Role of Surface Mining

The fact that European surface mining has made huge progress in improving its negative image as a noisy and dusty industry is something that is perhaps not always appreciated or understood.

It is inevitable that where there are earthmoving machines there will be noise and dust, but some mines – especially those in Britain – are going to great lengths to reduce levels beyond what is expected. Novel-noise and dust-suppression systems, such as those developed by Banks Mining and Miller Argent, are to be applauded for leading the industry in the right direction.

Safety is yet another area that has come under serious scrutiny. It is not just about wearing the right clothing and protective gear, safety also applies to machinery. There is no better example of what can be done to improve safety around machines than those measures

taken at Broken Cross, a coal mine owned by Scottish Coal. This mine has put a great deal of money and thought into raising safety awareness in and around their huge trucks.

However, despite the expensive efforts being made to increase the safety of workforces as well as reduce environmental impact, Britain's coal mines continue to attract protesters who are keen to shut them down. Perhaps the answer to this problem lies with the practices of other mines in Europe. British mining ought to consider three European mines, in particular; mines which not only bring more transparency to the business but also make a bit of extra cash in the process.

One of Europe's best kept secrets concerns New Boliden's Swedish Aitik copper mine, Rio Tinto Minerals' French Luzenac talc mine and RWE Power's huge

Tourists taking in the view of the Luzenac talc mine in the French Pyrenees.

A view of the German lignite mines owned by RWE Power.

German lignite mines. The secret is that all of these mines can be accessed by tourists at certain times of the year.

There is huge interest in all three locations, which attract over 120,000 visitors a year between them. While the big machines are the star attractions, the general public is also keen to learn more about new measures being taken to reduce noise and dust. Visitors also express an interest in learning about the restoration plans for the land once the mining is finished.

In light of this, I sometimes feel that we live in a hypocritical world. We all reap the rewards of surface mines everywhere and are all responsible for the ever-increasing demand for raw materials.

As our insatiable appetite for energy continues to increase, I applaud the fact that the there are 6,000 biogas installations in Germany that are currently generating the same volume of electricity as two nuclear power stations. Ultimately, however, I question whether this, too, will be sustainable in the future.

With a global population forecast to rise to 7.7 billion in 2020 and still further to 9.6 billion by 2050, we may not be able to afford the luxury of using land to generate energy. In the future, we may need every hectare of available land for food production.

Therefore, it is not a case of choosing between renewable and non-renewable energy sources. We will soon need to use every form of energy we can lay our hands on, including nuclear power, gas, coal, oil, solar, wind energy and any other new forms that may be developed in the future.

Currently, we are at a crossroads in the energy debate, with Brussels politicians controlling the future of the last remaining coal mines in Europe. The future is uncertain and some of the big machines in the state-subsidised Spanish coal mines have already paid the price when they were brought to a halt for several months.

In the worst-case scenario, it is not just Spanish coal mines that could be forced to close permanently within just a few years, but coal mines Europe-wide.

Aitik copper mine in Sweden runs tours of their facilities.

Despite these concerns, while the current economic climate is perhaps not the best time for coal-mining companies to invest several million pounds in new earthmoving machines, there are a few surface mines – especially those in Scandinavia which mine gold, copper, nickel and iron ore – that are clearly confident for the future.

This is especially true in the more remote regions of northern Europe where they are seeing current mining operations expanding and new mines opening up. Could this possibly have something to do with the fact that there is little resistance to expansion – or to the mines in general – in these sparsely populated areas?

Large Machines

These new operations need new machines and already several hundred million euros has been invested in more brand-new massive earthmovers for the Swedish Aitik copper mine and the Finnish Talvivaara nickel mine.

Scandinavia is the playground for the biggest Bucyrus and P&H electric rope shovels in Europe, not to mention the largest mechanical wheeled loaders and some of the biggest dump trucks in the world.

In terms of hydraulic excavators, Europe has the second largest offerings from all the major suppliers. However, we can take pride in the fact that three of the largest and most powerful excavators in the world – the Liebherr R 9800, the Komatsu PC8000 and the Bucyrus RH400 – are all constructed in factories located in the heart of Europe.

With the exception of a couple of sites in Scandinavia and Spain, there are few European mines that can justify the services of any of the truly massive mining monsters. However, now that costs per tonne are beginning to play an even greater role, only a fool would rule out the possibility of any of these monstrous excavators working in Europe some time in the future.

At the time of completing this book, Caterpillar's multi-million dollar bid for the Bucyrus empire was still awaiting approval from the regulatory bodies. Assuming

Scandinavia is the playground for the biggest P&H electric rope shovels in Europe.

that this mega deal does go through, Cat may soon have a very large product portfolio. While monster electric mining shovels and huge dump trucks are low on the list of all but a few of the largest mines in Europe, the RH series excavators are another matter. The RH120, in particular, is hugely popular with British mines.

After Terex acquired the former O&K excavator range, there were grumbles about the difficulty of sourcing parts and since Bucyrus took over, they have worked hard to address their customers' concerns.

There is no doubt that in Britain these mutterings have opened the door for the competition. Komatsu is the most successful and is taking full advantage of the situation, while Liebherr is doing its best to spoil the popularity of RH excavators by cultivating a firm foothold in Scotland. Hitachi has even stepped onto the scene and managed to gain a toehold on the Welsh mining market.

However, all these new contenders still have to prove their reliability and longevity. If Caterpillar gets the green light to forge ahead with its takeover of Bucyrus, then it could be just a matter of time before the RH excavators are offered with service and back-up that is second to none. (See Postscript.)

CHAPTER 1

Italy

Dump trucks from well-known suppliers like Caterpillar, Komatsu and Terex may be the preferred option for many of Europe's largest quarries and mines. In Italy, however, there is another household name which has long proved to be a worthy competitor. The company is Perlini, a comparatively small, family-owned business with two factory locations near Verona, which has produced more than 10,000 trucks over the last fifty years!

CEO Roberto Perlini founded the business in 1957 and although present-day operations are managed by his sons, Francesco and twin brother Maurizio, Roberto is still very much in control of the family-run empire. None of the Perlini family members are under any illusion that they will ever rattle the cages of the larger, more established dump truck producers. 'We compare ourselves to a small, family-owned pizza company having to battle against the likes of Pizza Hut,' jokes Francesco.

Underneath this light-hearted comment, however, business is serious, especially since Perlini claims to make one of the best dump trucks on the market. The company also lays claim to a number of technical firsts, including the first to include an electronic engine and transmission in a truck in 1986 and the first to use a Detroit Diesel engine. Unfortunately, while the brand is well established in many parts of the world, Perlini has never managed to get a foothold in the US, a fact that Francesco puts down to the stronghold that the big names have over that particular dump truck market.

The story is similar in the UK. The only two machines ever sold to Britain were fifteen years ago and these are no longer active. The company is not overly confident in its ability to break into the American or British markets and Perlini is not prepared to lose money trying.

1.1. Perlini dump trucks are a common sight in Italy. This row is waiting in line at a Holcim quarry.

So Italy remains the company's stronghold, while China is one of the biggest export markets. Obtaining exact numbers is difficult because it is such a big country, but the active park is estimated to be above 2,500 units. Business in China was boosted during the 1980s and early 1990s by major orders for trucks to work on new dam projects. These orders included 400 trucks for the Gezhouba Dam in 1980 as well as 600 dump trucks – the single largest order Perlini has ever recorded – just a few years later!

Turkey is another important export market for Perlini. The largest recorded Turkish order was for eighty-three trucks in 1983. The trucks were used in the construction of the second bridge over the Bosporus near Istanbul. Another seventy trucks were also purchased to help with the construction of a Turkish motorway. However, there is no need to travel that far to see a Perlini dump truck, because there are plenty of examples of these home-grown machines working the landscape of Italy.

The Italcementi cement group runs a fleet of fifty Perlini trucks at its Italian quarries. Another important customer, which is also in the cement business, is Holcim and they own and operate a number of trucks. Finally, Riva Steel, the second largest steel giant in Europe, runs a fleet of one hundred Perlini trucks in Southern Italy.

Perlini currently makes a range of four models of DP dump trucks that are available for purchase. They are not the biggest in the business but the top-of-the-range DP 905, which has a rated payload of 95 tonnes, is definitely a sizeable hauler. In fact, with the exception of the engines and transmissions, Perlini makes eighty per cent of the truck components in-house. This includes the axles, tipping body, hydraulic cylinders, axle suspension systems, disc brakes and even the cab. Without bringing colouring into the equation, to the untrained eye one dump truck looks very much like another so at first it is difficult to see where Perlini may have the upper hand. Their trucks only have a limited number of options, which include a 30 or 35 mm thick steel base for the tipping body (standard 25 mm), a central lubrication system, larger as well as thicker profile tyres and the usual comfort additions like air-conditioning and radio. Also, the standard dry disc brake system fitted to the DP 705 can be swapped for a wet disc brake system. 'Ninety-seven per cent of our customers only require a few options,' says Francesco. 'We prefer to invest our money in hi-tech, quality solutions rather than gimmicks.'

Perlini's biggest single strength could well be the ability to be flexible. 'Being a small company means that we can change our focus quickly when it becomes necessary. Our current spearhead involves new axle suspension, engine and transmission technology but if the market changes, we can adapt more quickly than larger companies.'

One further benefit of being small is being able to make customers feel important. 'When I am dealing with customers I tell them to contact me directly if they need help. I also make a point of going to see individual customers if there are problems. Being small makes us accessible and customer friendly.'

However, there are plenty of drawbacks to being a small company among giants. He admits that it can be difficult to always provide the best service in the field. 'This is an area we need to improve but it would be unrealistic to think that we will ever be on the same level as the major players. It takes a strong company to maintain service and customer relations.'

Biggest Perlini Hauler

One of the Holcim cement group's quarries near Milan runs seven Perlini trucks, one of which is a DP 905. Each year the quarry extracts nearly 1.5 million tonnes of limestone. The Perlini trucks operate for one 10-hour shift per day, from Monday to Friday, and are filled by a fleet of Cat machines.

The power behind the DP 905 comes from a 1,050 hp engine. When loaded to full capacity, the Italian-made heavyweight tips the scales at nearly 160 tonnes. Although the DP 905 is not as large as many of the competitors' trucks, Francesco says that they often have trouble tempting operators to move up the power scale from one of their three smaller trucks. 'The smaller trucks are normal quarry trucks. Compared with the DP 905, they are almost toys; it is a real machine!' The bigger dump truck is also in a totally different category when it comes to maintenance. 'It takes about an hour to change the tyres on the smaller trucks, whereas changing the tyres on the DP 905 takes half a day!'

One of the big advantages of the extra capacity offered by a truck with a 95 tonne payload is the opportunity to use it to replace two smaller trucks. 'It offers the same production as the two machines because loading and unloading only has to be done once. Plus, it eliminates the need for another driver.'

1.2. When fully loaded the DP 905 carries 95 tonnes of material making it a good deal larger than the DP 705.

The downside to operating only one truck is that when it breaks down everything grinds to a halt. In seeking additional security against this possibility, many Perlini customers opt for a service contract. 'Customers pay a fixed price per hour and we guarantee to keep the vehicle working at all times.' Perlini service contracts are very popular and nearly ninety-five per cent of the trucks working in the cement sector in Italy have been entered into the scheme.

The Perlini portfolio is an impressive one. During the 1980s and 1990s, often referred to as the company's heyday, Perlini made an average of 600 dump trucks a year. Unfortunately, production has now slumped to less than a hundred trucks a year. The majority of these remain in Italy, while quite a few go to Spain. Roberto knows that a hundred trucks a year is simply not enough to keep the company in the black but, though there is plenty of factory space to produce between 150 and 170 trucks a year, the declining sales will not support an increase in production. 'A small business like ours does not find it easy to go head to head with

the major brands. Production is down at the moment but we are confident that it will pick up in the future.'

As for the potential for future sales closer to home, Roberto is concerned that the difficulty of obtaining new extraction licences for quarries in Italy will spread to their customers in Germany, Spain and France. The Italian government tends to be a bit left wing and the country also has a powerful environmental movement, neither of which is particularly sympathetic to the quarry industry. Quarries are allowed to remain in operation but most find it almost impossible to expand.

Roberto thinks that the business depends on whether or not extraction licences will be forthcoming in the future. 'If so, customers may be prepared to invest in new dump trucks and if not, then they will attempt to keep existing machines in operation as long as possible.' He foresees some customers still trying to work with machines that have clocked up over 75,000 hours! 'Dump trucks are like cars: a twenty-year-old Mercedes will not run as smoothly as a new model.'

1.3. This DPT 70 is a brand-new off-road hauler from Perlini that was launched at the 2010 Bauma show. A strange mix between a rigid truck and an ADT, the newcomer has a 70 tonne payload and is powered by a 550 hp MTU Detroit Diesel engine.

Working in White

At the Monte Valerio quarry, two hours north of Rome and within sight of the Mediterranean coast, two 65 tonne Perlini trucks are the favoured dump truck choice. The site is home to one of the largest limestone quarries in Italy. Nestled at the top of a ridge overlooking the village of Venturina, it surely rates as one of the most attractive quarries in the country. On a clear day the view from the top is quite remarkable and sightseers are easily able to see the island of Corsica.

In the business of extracting nearly 700,000 tonnes of limestone a year, the quarry is owned by the Sales construction company. Their site has drawn attention not only because it is home to Perlini trucks but because Monte Valerio owns a stunning 120 tonne excavator. The R 984C is not the largest Liebherr excavator in Europe but the truly magnificent setting of the Monte Valerio quarry is a fabulous place to see one in the flesh. Besides, the excavator was one of the first R 984Cs

commissioned in Europe.

The prime mover at this quarry used to be a 455 hp Cat 5080, an old machine that was made in 1996. After digging non-stop for 13,000 hours, however, the old Cat was replaced by an excavator that was more modern and powerful as part of a plan to double future production outputs.

Production manager Milco Bolognesi says that management looked long and hard at a number of different candidates, including a Cat 385 and a Hitachi EX1200, before making a trip to Germany to see a Liebherr R 984C. The Cat and Liebherr excavators were shortlisted but, in the end, the Liebherr excavator was thought to be better suited to the quarry. 'To increase production, we need an excavator that can handle hard work and we are under the impression that Liebherr machines are more reliable,' says Milco. 'It just looked a stronger machine.'

In a moment of vanity, the deal was clinched by the colour scheme. The white colouring on the Liebherr matches perfectly with the livery of the two Perlini haulers. 'They look great together. It is a perfect marriage.' Under the brightness of the Italian sun it is difficult to argue with him: the limestone face provides a fantastic backdrop and the two machines look quite spectacular.

Speaking of spectacular, the output of this substantial excavator is just that. The Liebherr is fitted with a 7.8 m boom, which is the smallest of three options. Connected to a 3.4 m dipper and a standard, 7 cu m rock bucket, the machine is easily capable of sorting and shifting large blocks, some weighing as much as 8 tonnes. The 710 hp engine also provides plenty of power for the bucket teeth to prise apart smaller blocks.

The R 984C is made at a factory in Colmar, France. Typically, big machines tend to be sent in bits and pieces from factories but this Liebherr model is the largest excavator ever to leave the factory as a complete machine. Though the R 984C is technically one of the smallest mining excavators made by Liebherr, it still lifts nearly 11 tonnes of material at a time and is capable of filling the two Perlini trucks in only six attempts. During a normal nine-hour stint this excavator shifts around 4,000 tonnes of material.

1.4. At this quarry, a Liebherr R 984C excavator is the centre of attention, instead of the Perlini trucks.

The cab is 800 mm higher than a standard one. Operators like this design because it allows them to look right inside the trucks. In fact, the previous prime mover, the Cat 5080B, also had an elevated cab. As well as providing better operator visibility, the raised cab has another benefit for quarries like Monte Valerio. 'All rock

1.5. During a normal working day the R 984C can shift as much a 4,000 tonnes of material.
This machine weighs nearly 120 tonnes.

1.6. It takes a big excavator to dwarf this Astra road truck.

1.7. The Liebherr excavator has no problem lifting blocks weighing 7 or 8 tonnes.

at this site is blasted, which can occasionally result in overburden piles over 10 m tall. The high cab position comes in handy in such scenarios.'

The most visible differences between the outgoing Cat 5080B and new Liebherr R 984C can be noted at the front. The Cat was a front shovel, while the newcomer has a backhoe arm. There is good reason for the change. 'Our operators say that a backhoe offers higher levels of digging performance than a front shovel attachment. Backhoes provide faster, more efficient loading and are better suited for loading big rocks. The extra reach also helps.'

Relegated from its prime mover role, the engine on the Cat 5080B has been overhauled but the excavator is really only used when the Liebherr requires a service. 'The Cat is useful as a back-up,' says Milco. 'But at some point there will be a major problem and we will have to get rid of it.'

Baby of the Fleet

Hitachi's giant EX excavator series boasts a colossal, top-end 800 tonne digger. Unfortunately, there are none of these in Europe. However, there are small numbers of the rest of the range, all of which are covered in later chapters.

The EX1200, an excavator that weighs around 110 tonnes, is the baby of the EX range but is believed to be the largest EX excavator in Italy. Officially launched in 2002 as the successor to the EX1100-3, it is a machine that gradually evolved into the EX1200-5D. More recently, the EX1200-5D version was superseded by the EX1200-6. However, the EX1200-5D remains a popular machine and can be found at large quarries and mines across Europe. There are quite a few in Italy. One, in particular, works in the rugged, sparsely populated area of Tuscany at the Marna limestone quarry, which is located not far from the village of Rassina, an hour's

1.8. The Marna quarry has two Perlini DP 705 dump trucks, which were purchased to replace four smaller versions.

1.9. The DP 705 is the second largest truck made by Perlini.
It has a payload of 65 tonnes.

drive south of Florence. One of the biggest quarries in central Italy, Marna is owned by cement producer Colecem, while all overburden and limestone is extracted by contractor Casentino Scave.

Marna production manager Antonio Girei explains that the Hitachi excavator was bought as a cost-cutting exercise to replace two smaller excavators. The contractor is a valued Perlini customer. At one time they used to operate four 40 tonne capacity DP 405 trucks with the two smaller excavators. When the EX1200 came in, these four smaller trucks were swapped for two larger DP 705s, each with a 65 tonne capacity. 'We ditched an excavator and two dump trucks,' says Antonio proudly. 'But we still managed to achieve the same production outputs with the added bonus of reduced production costs.'

This EX1200-5D does all the major quarry work alone. The excavator is equally at home loading overburden at the top of the quarry as it is filling trucks with limestone at the bottom. Lifting nearly 10 tonnes at a time, in a good dig the 5.6 cu m rock bucket is capable of filling one truck in just a few minutes.

The production manager confirms that the Hitachi excavator has met all his expectations. 'The excavator is easily capable

1.10. The EX1200 is the smallest model in Hitachi's EX series. This Dash 5 works at the Marna limestone quarry.

1.11. The Hitachi excavator is capable of filling this 65 tonne capacity Perlini dump truck in five or six attempts.

1.12. The 5.6 cu m rock bucket easily empties ten tonnes of material at a time into the Perlini truck.

of producing the same output as the two smaller excavators.' During the first year of operation it shifted over one million tonnes of material and is known to regularly handle around 7,000 to 8,000 tonnes of material per day.

The EX1200-5D is a powerful workhorse but the quarry has limestone reserves that have been estimated to last for at least twenty to twenty-five years and it is almost certain that the excavator will not last that long. Monte Valerio is already contemplating an even more powerful excavator when the time comes to change. Ultimately, management would like to swap their dump trucks for larger 95 tonne versions as well. 'Our expectations for the future are high. We have already managed to cut the number of staff and machines by fifty per cent, which is enough for now.'

CHAPTER 2

The Netherlands

Although there are no mines in the Netherlands, there is a quarry and quite a big one at that. It is located deep in the southern province of Limburg, a region best known for its valuable asparagus crop. This is a landscape that is home to rocky outcrops which ultimately rise to become the Belgian Ardennes.

The St Pietersberg quarry – the largest in the country – is concealed on the southern side of the hill of the same name; indeed it is so well hidden that touring cyclists riding alongside the River Maas often pass by the entrance without noticing it.

On the northern side of the St Pietersberg hill is Maastricht, the oldest city in the Netherlands. Many of its older houses and churches were built from large blocks of limestone carved out by hand from the Maastricht side of the hill and transported by horse and cart. Although major excavation work stopped fifteen years ago, some smaller quarries still excavate the limestone for restoration work.

A hundred years of excavation has left the hill riddled with more than 20,000 passageways and a network of caverns that run for more than 150 km. During World War II these caverns were used as hiding places for locals seeking shelter. In fact, the caves to the north of Maastricht were also used to hide several hundred world-renowned works of art, including Rembrandt and Vermeer classics. The Germans never did find them.

The St Pietersberg quarry operation now belongs to ENCI – the First Dutch Cement Industry. The deep layers of high-quality limestone on the site are the basis of more than a million tonnes of Portland cement a year.

Present-day excavations have exposed many of the underground caverns on the south side of the city.

2.1. One side of the quarry is completely riddled with caves that run for many miles underneath the ground. Evidence of limestone quarrying in this area has been found to date back over a hundred years.

2.2. ENCI cement is famous throughout the Netherlands. Each year vehicles travel the length and breadth of the country carrying nearly a million tonnes of cement.

Maastricht's tourist industry has taken advantage of the discovery of these preserved caverns and now runs guided tours into what are now known as the St Pieter caves. Many of the quarry's current staff recall playing in the maze of tunnels during their childhood, but the extensive labyrinth is now banned as a kids' playground. Theatrical groups still sometimes use the caves as a dramatic venue for evening performances; otherwise, they are now home to large numbers of bats and a pair of very rare birds. These are Eagle owls, Europe's largest owl, and they are the only breeding pair in the Netherlands. The pair have made one of the cave entrances their home and, as a result, ENCI has abandoned that end of the quarry in order to allow them to nest in peace.

Much has changed during the past fifty years of limestone extraction. Back in 1965 there were nearly 1,200 staff employed at St Pietersberg quarry. Today this number has been slashed many times over. Despite large production volumes and guaranteed limestone reserves for years to come, the future of the country's only quarry hung in the balance for many years.

In 2005, parent company the Heidelberg Group announced plans to close it within three years, citing high production costs as the main reason. In the end, they agreed to a five-year stay of execution and an additional restructuring programme to reduce operating costs. Staff were relieved, but this extension still resulted in the loss of 140 of 320 jobs.

Once the decision had been made to keep the quarry open, the next major headache was determining the best way to revamp the ageing machine fleet for the least amount of money. By Dutch standards, the four trucks that they had in operation were large; they had one 45 tonne capacity Cat 773B from 1990 and three 60 tonne 775Bs (two from 1993 and one from 1995). All these machines had notched up over 25,000 hours of work. The mine's Cat 988F wheeled loader was also a concern. Built in 1994, it had also operated for over 25,000 hours while working at St Pietersberg.

After careful consideration of the options, ENCI decided to put three of their oldest trucks and the wheeled loader through the Cat Certified Rebuild (CCR) programme. One by one Dutch Cat dealer Pon Equipment collected the four machines and after two days of thorough cleaning, each truck was stripped into thousands of different pieces. All metal components, including the front and rear frames, cab frame and

2.3. *Three of the four Cat trucks were put through the Cat Certified Rebuild programme. Each one has now clocked another 9,000 hours and will soon need replacing.*

wheel rims were sand-blasted back to the bare metal. All hoses and electric wires and cables were stripped. Then, all main components – such as pumps, engines and drivelines – were thoroughly tested. Cab interiors were also stripped to the bare frame and, following that, the seats – as well as all upholstery, all controls and glass panes and seals – were replaced.

Three months and nearly 8,000 new parts later, each truck was fully assembled and ready to start work. They were returned to the quarry in practically new condition and with full warranty. Although it cost ENCI more than half the price of a new truck, quarry staff were amazed when they were returned. 'We knew they were the same trucks but they were unrecognisable,' says Jo Kleijnen, who was the head of the quarry's technical department at that time. 'They looked brand new. They had new serial numbers and even their tachometers were turned back to zero. It is such a shame that within two days of operation they were once again caked in limestone.'

All three trucks have since gone on to acquire another 9,000 hours of operation but the present-day management admit that they would probably not use the restoration programme again. Unfortunately, the Cat

trucks are coming to the end of their working lives. The truck that was not put through the programme has done over 27,000 hours and the other three have clearly seen better days. 'They are not falling apart but we cannot keep pumping money into them,' says Fred van Bijnen, the current head of the technical department.

In order to solve the problem of their ageing fleet, a Heidelberg Group expert looked at the best machinery options for their quarries. After comparing a number of 60 tonne capacity trucks from providers such as Cat, Hitachi and Komatsu, the expert came to the conclusion that the Cat trucks would be the best choice, due to their low fuel consumption and greater power. However, while the expert advised Cat replacements, the company's recent purchase of a brand-new Komatsu HD605 seemed to confirm a wind of change.

Nearly 1.8 million tonnes of limestone has to be extracted every year to produce the annual quota of cement. In order to achieve this quota the quarry requires the combined efforts of a pair of substantial diggers. Twenty years ago the St Pietersberg quarry relied on a Liebherr R 984 excavator and an electrically powered Ruston rope shovel but nowadays these machines have fallen into disrepair. When the track

2.4. *The quarry's Cat 988F wheeled loader has notched up another 12,000 hours since it was rebuilt. Here it is shifting crushed flint.*

2.5. *The St Pietersberg quarry recently acquired this new Komatsu truck with a 60 tonne capacity.*

2.6. This Ruston-Bucyrus 150-RB rope shovel once needed 20 staff to help shift the electricity supply cables when it had to change locations. Unfortunately, this machine is no longer active.

tensioning system eventually broke on the Liebherr, there was still work for it as a static machine and for several years it loaded flint onto a transport conveyor at the far end of the quarry. Unfortunately, it was recently sold for scrap. The Ruston-Bucyrus 150-RB rope shovel now stands idle on a specially made ramp near the main oven. It remains to be seen what will happen to it when the quarry closes.

More recently, two Liebherr R 994 excavators took on all the limestone extraction in order to contribute to the production of cement. Weighing nearly 230 tonnes each, these machines are the largest Liebherr excavators in the Benelux – Belgium, Netherlands and Luxembourg. The first one was bought new in 1993 and fitted with a 6.5 cu m backhoe bucket. During the past twenty years it has seen a great deal of action – almost 36,500 hours. However, it too, now stands idle. With badly stretched tracks and over 20,000 hours on the clock the 1,273 hp engine has had it. Management at the quarry are currently deliberating its future: the choice is either to sell it, strip it for donor parts to keep the second identical machine running or spend nearly €200,000 to renovate it.

The second Liebherr R 994, a face shovel, was sourced from a German quarry in 2001. Shipped in crates, it was put together and ready for action by the summer of 2002 but the 13.5 cu m bucket it came with proved too large to penetrate the tough limestone. The problem is that the quarry has two main layers: a softer yellow top layer of Maastricht limestone and a tougher bottom layer of Gulpen limestone. The Gulpen limestone, in particular, is famous in Limburg because it shares its name with the River Gulp and Gulpener beer. The quarry needs to mix both types together in order to obtain the best quality cement. As this Liebherr was intended for use on the tougher Gulpen limestone, the previous 13.5 cu m bucket had to be swapped for a new and smaller 6.5 cu m one, which when full, holds 11 tonnes.

When ENCI first acquired the 2001 Liebherr R 994 the machine had done 12,000 hours; today, it is pushing 30,000. Due to its age, it spends more time in the easier dig near the top of the quarry. The rear counterweight is now showing signs of cracking at the mounting point

2.7. After nearly 20 years of action the backhoe version of the quarry's Liebherr R 994 has had it. It will probably be scrapped.

2.8. The ageing Liebherr R 994 front shovel excavator spends most of its time up in the softer yellow Maastricht limestone at the top of the quarry.

2.9. The quarry comprises two different layers of limestone: the yellow Maastricht limestone at the top and the hard grey Gulpen limestone underneath.

and there are plans to swap this for the ballast on the static machine, which suggests that the decision to sacrifice at least one of the excavators has already been made.

The St Pietersberg quarry has faced many uncertainties over the years but the final decision on its future has been made – limestone extraction must end in July 2018. An additional year has been earmarked to restore the site, during which time ENCI will still be allowed to use the oven for a maximum of ninety days to produce a limited amount of cement. After July 2019, however, limestone quarrying in the Netherlands will be history. While the decision is a sad one, at least ENCI is now able to plan for the final few years. The parent company, Heidelberg, is not keen to spend any more money, but some form of investment on the hauling front will be unavoidable in order to complete the job.

Biggest in the Benelux

Some cash was recently earmarked in order to buy a replacement excavator to supersede the redundant Liebherr R 994. Quarry operators made no secret that they wanted to replace it with a newer model of the same machine. However, Liebherr recently updated its

mining excavators: the R 994 was phased out to make way for the R 9250, while the heavier R 994B was ousted for the R 9350. Liebherr has two production sites – one in Colmar in France and the other in Brazil. When ENCI was in the market for a new excavator, the French facility had sold out of R 994s in preparation for the first batch of R 9250s. The Brazilian factory, however, was still assembling the last two R 994Bs and, as the Dutch quarry was keen to obtain one of them, it took out an option. Unfortunately, a Russian customer bought both of the 994Bs before the deal was concluded, causing a problem for the St Pietersberg quarry.

They continued to seek a replacement. In the meantime, several staff members took the controls of a number of large Terex O&K machines in Germany but secretly, they hoped for a brand-new Liebherr R 9250. This was not to be; in the end, Komatsu clinched the deal with its PC3000 model. The parts of the 260 tonne Komatsu excavator, the largest in the country, started to arrive on site in June 2009. The full details of the deal between the Heidelberg Group and Komatsu will probably never be disclosed, though it is quite likely that the decision to buy Komatsu was, in the end, a financial one. The Heidelberg Group asked several

2.10. The Komatsu PC3000 is the largest hydraulic excavator in the Benelux. During its first year of operation it did 2,000 hours.

suppliers to provide them with quotes for the cost of more than a hundred new machines for its global operations. The assumption is that Komatsu gave them the best deal.

During its first year of operation the PC3000 worked for nearly 2,000 hours. It spent most of its time in front of the harder benches, while the easy digging in the Maastricht limestone was reserved for the sole surviving Liebherr. Working for two 8-hour shifts each day, from Monday to Friday, these two excavators needed to stockpile 12,000 tonnes a day, or enough limestone to keep the cement plant working over the weekend. The Komatsu excavator did and still does the lion's share of the work – around 9,000 tonnes a day – but on a good day the old Liebherr can still shift 4,000 to 5,000 tonnes.

The ferocious power of the newly acquired PC3000 has made a positive impact not only in the obvious sense of speeding up the extraction process but also in easing conflict between ENCI and the inhabitants of the nearby town of Maastricht. Noise pollution is one of the most common complaints that ENCI has to combat and the powerful PC3000 has helped to alleviate those

complaints with its ability to extract even the toughest layers of limestone without the need for drilling and blasting. Used only as a last resort, the drill rig now stands idle in the workshop. 'That is unless one of the two excavators is out of action,' says Theo Kicken, who headed up the afternoon shift. If this scenario occurs, the Cat 988F, which is fitted with high-lift loading arms, is called on to load the trucks. 'But the Cat 988F can only operate in this way when the limestone has been blasted,' he says. The quarry used to operate in three shifts but noise-reduction laws mean that they can no longer work at night.

Noise complaints aside, the quarry has an even bigger problem to deal with – flint. The limestone layers are full of the stuff: it is unbelievable just how much can be found on the St Pietersberg site. The reason that flint is a problem for the quarry is because it is incredibly sharp and wreaks havoc on truck tyres, easily able to slice off complete truck tyre lugs. The machine's bulging rubber side walls offer little resistance to the flint's cutting power and flint lumps flung out from between tyre lugs damage truck oil tanks, requiring new tank plates to be fitted regularly. Excavator and wheeled loader buckets also

2.11. The new 260 tonne heavyweight makes light work of digging up even the toughest limestone.

suffer. The leading edge of the 988F bucket generally needs repairing five to six times a year and in a thick flint layer excavator bucket teeth wear in just a few months.

The fact is that the St Pietersberg quarry could do more to limit the damage to their fleet but the recent round of cost-cutting measures means mechanical funds are in short supply. It is no secret that the condition of mine and quarry roads is paramount to prolonging truck tyre life. The quarry's initial solution to this problem was a Cat scraper that patrolled the haul roads.

Unfortunately, it would have cost a considerable amount of money to replace this ageing machine so the company went for a cheaper alternative and bought a new and modified 230 hp New Holland tractor fitted with a mid-mounted blade. Unfortunately, this cost-cutting measure proved largely ineffective as some of the lumps of flint were so big that they stopped the tractor.

Despite its incredible power, the arrival of the new Komatsu PC3000 excavator was still met with mixed reactions. Operators commonly become attached to their machines, and, as a result, changing brands can often be a difficult and touchy subject. ENCI's Liebherr excavators may have seen their best days, but their

operators loved them and were not keen on a change. This fact is best illustrated by the story of a former excavator operator who had always worked with the Liebherr R 994. As the story goes, no one was allowed to touch his machine unless he had a day off or was sick and he vehemently vowed he would never be seen dead on a Komatsu. True to his word he never did; he died of a massive heart attack soon after the PC3000 was delivered and staff now joke it was the final nail in his coffin!

The true test of the new excavator and the moment that cemented it forever in the hearts of the quarrymen, was when it went head to head with the R 994 at the bottom of the pit. Were they secretly hoping it would fail the test? Both excavators had to dig in the tough Gulpen limestone and, for the occasion, the 6.5 cu m bucket attached to the Liebherr was equipped with special long teeth. These teeth snapped off quickly and the excavator was unable to gouge out a full bucket, while one of the teeth on the Komatsu's bigger 8.5 cu m bucket also broke but it had no trouble scooping up 14 tonnes at a time. It is now fitted with shorter and blunter teeth, and the quarry staff are still just as

2.12. The limestone layers contain large quantities of flint, which can wreck wheeled loader buckets and slice truck tyres to shreds.

2.13. This New Holland tractor has been specially modified to grade the quarry roads but it struggles with large flint lumps.

impressed by its power as they were on that first test day. 'I love it,' says operator Marc Gulikers. 'It is far more powerful than the Liebherr R 994, but the Komatsu is twenty years younger and I would have liked to see a new Liebherr R 9250 work on the same spot.'

Before the switchover, Marc spent nearly a decade at the controls of the two Liebherr excavators. He used to work as a postman but after eleven years at the wheel of a post van, he was ready for something completely different. Marc started a café in a nearby town but in 1997 when things did not work out, he took a job at the St Pietersberg quarry as a truck driver. He soon fell in love with the excavators, and after completing a training course, has never looked back. Marc says he relishes the ferocious appetite of the PC3000 but he still prefers the Liebherrs. 'The Komatsu is a totally different animal,' he says. 'It contains so many safety mechanisms that sometimes it just shuts down and there is nothing you can do about it.'

Marc thinks of the Komatsu as a nervous machine, while he considers the Liebherr to be much easier to control. As an example, the quarry works with a bench height of around five metres and the backhoe excavators stand above this. Large amounts of flint will cause the Liebherr to grind to a halt, which happens less frequently with the Komatsu; however, care is needed with the Komatsu in other operations. 'The Komatsu is so powerful that the bucket sometimes tries to pull the excavator over the edge,' Marc explains. It also took him a while to get used to something as simple as stopping the excavator. When the joysticks on the Liebherr are released the machine stops immediately, while the PC3000 tends to continue to run for a bit. 'This was unnerving in the beginning because I thought I was going to topple over the edge of the bench, but now I am used to it.'

Another difference between the Liebherr and the new Komatsu is the size of the cab. The cab on the Liebherr excavator is huge; there is ample room to swivel the trainer seat alongside the operator. The 'smaller' cab on the Komatsu is seen as a big disadvantage because it has less space and smaller windows, meaning less glass, so reduced visibility is a constant problem for operators. Marc says, 'I sometimes have to strain to see whether there are any trucks behind the excavator, which is something I never had to do in the R 994.' Some

2.14. PC3000 operator Marc Gulikers used to deliver post but today he is equally at home at the controls of a big hydraulic excavator or behind the wheel of one of the trucks.

operators also complain that the Komatsu's larger boom restricts visibility across the top when loading a truck to the right. The design is smaller on the Liebherr so drivers can see across the top. Not all the complaints are directed at the Komatsu, however. One problem that operators have highlighted about the old Liebherrs is that they are incredibly noisy in the cab and all their operators have to wear earplugs. On the other hand the Komatsu cab is equipped with noise-reducing mechanisms.

There are certainly a couple of things in the Komatsu's favour but there is one thing none of the excavator operators will ever get used to – the starting performance of the Komatsu excavator on a cold winter morning. After a night of minus ten degrees Celsius, a temperature that is not uncommon in the Netherlands during the winter, Fred, current head of the technical department, reckons it takes up to forty-five minutes for the excavator to reach a reasonable operating temperature. 'It has difficulty on a cold morning,' he says. 'Even at temperatures of minus four to minus five degrees Celsius, the Komatsu can take nearly half an hour to warm up fully .'

Whether the pros outweigh the cons or not, the St Pietersberg excavator operators are stuck with the Komatsu because it will easily outlive the current project. There are already plans to move the German-made machine back to its homeland and to another of the Heidelberg Group's quarries once St Pietersberg closes.

Maasvlakte 2

What the Netherlands lacks in quarries and mines, it more than makes up for with massive land-winning reclamation projects and there is no better place to witness this than the current expansion work in the Port of Rotterdam.

The site is called Maasvlakte 2. When finished in 2033, the €3 billion investment by the Port of Rotterdam Authority will provide an additional 1,000 hectares of industrial area and boost container facilities by twenty per cent.

Every year more than 34,000 seagoing ships and another 133,000 smaller vessels bring 100 million tonnes of crude oil and 365,000 new cars to the Dutch port. Rotterdam also enjoys the status of being the largest container port in Europe – when laid end to end the 10 million containers handled each year could easily circulate the globe.

Forecasts show that volumes at Rotterdam will rise

2.15. When Maasvlakte 2 is finished – which will be around 2033 – the site will provide the Port of Rotterdam Authority with an additional 1,000 hectares of industrial area. It will boost container facilities by twenty per cent.

2.16. Nearly 365 million cu m of sand needs to be pumped ashore from the North Sea to form the basis of the new land area.

sharply during the next few years. The number of containers alone is tipped to rise to 17 million by 2014; then 23 million by 2025; and ultimately to 38 million by 2035. The fact is that the sprawling complex is already fully stretched by current volumes, so the only way to meet this growth is to further expand the port. Almost 700 hectares of newly reclaimed land has thus far been earmarked for container handling, which will be sufficient to process the forecasted 17 million containers in 2014.

The land reclamation project will be carried out in stages. The first stage is to dredge sand stretching 12 km out into the North Sea. The sand works shot to national fame recently following the first recorded discovery of fossilised hyena droppings not far from the Hook of Holland. Research in the Natural History Museum Rotterdam has shown that the coprolite was produced by a cave hyena (*Crocuta crocuta spelaea*) some 30,000 to 40,000 years ago. Never before has such a young and nicely fossilised specimen from the late Pleistocene era been found in the North Sea basin.

They may yet uncover more startling finds because to

complete the new harbour nearly 365 million cu m of sand needs to be pumped ashore, which is the equivalent of filling the Feyenoord football stadium completely to the top nearly 250 times.

For the first two years, twenty-two huge trailing suction hopper dredgers shifted a staggering 170 million cu m of sand out of the sea. At the peak of the operation, nearly a dozen hopper dredgers pumped 3.8 million cu m ashore each week – a world record at a single job site. With men and machines working twenty-four hours a day and seven days a week, the project is on target to reach the 240 million cu m that is required to finish the first stage. The completion of this stage will coincide with the first container terminal receiving the first vessel in 2013.

Pumping sand to the shore is not the only major task because a huge fleet of big excavators, dozers and ADTs are involved in shifting the material once it reaches land. The Port of Rotterdam has built a visitor centre, called FutureLand, right on the edge of the project, meaning

2.17. Maasvlakte 2 is clear proof that the Dutch are masters at reclaiming land from the sea.
A special information centre called FutureLand has been set up to allow visitors to follow the project.

2.18. FutureLand provides visitors with an ideal opportunity to catch up on all the latest project construction films
and pictures as well as prehistoric findings.

that the sand works are enjoying plenty of interest. As well as affording good views over part of the newly reclaimed land area, the centre provides visitors with an ideal opportunity to catch up on all the latest news on the project, not to mention all the prehistoric findings.

There are also daily guided tours around the edge of the land-building activities. These tours are roughly an hour long and cost €5; they are hugely popular and tend to sell out weeks in advance. The country's largest earthmoving project is also one of the biggest tourist attractions. Covering nearly 2,000 hectares, Maasvlakte 2 will be the same size as Amsterdam's Schiphol Airport and Disneyland Paris.

Blockbuster

The stone side of the operation and the building of the new sea defences are seen as the biggest challenges of the entire project. The 11 km barrier that is being constructed contains a 3.5 km hard sea defence that will keep a batch of very special machines busy for the next two years.

Although sand makes up the centre of the 7.5 km soft sea defence, a phenomenal 5 million tonnes of rock is needed to part-build the hard sea defence. The rock used in this project arrives by boat at a rate of around 100,000 tonnes a week; most of it comes from quarries in Norway, Scotland, Belgium and Germany.

Even more impressive is that 21,000 square concrete blocks – each measuring 2.5 m – and an estimated 2 million tonnes of stone from the old hard sea defence have been salvaged to create the new one. Years ago, these older blocks were simply tipped into the water where they lie on the seabed at different angles. It took nearly a year and the combined efforts of a massive marine dredger and sophisticated electronics equipment to painstakingly extract just a handful of these blocks an hour.

Once salvaged, these 40 to 45 tonne blocks are loaded onto pontoons, transported back to dry land and then lifted off by massive 300 tonne cranes where they were

2.19. A total of 200,000 to 300,000 sq m of soft sea defences are being constructed by planting around 1.5 million Marram grass plants. This mammoth task is not expected to finish until the autumn.

2.20. The building of the new hard sea defence will require a phenomenal 5 million tonnes of rock and aggregates.

moved to a stockpile by a batch of unusual machines called Hover-Tracks. These Dutch-made creations feature a hydraulic side-shift mechanism to hold blocks in place, after which the complete structure rises clear of the ground.

The whole Maasvlakte 2 land-reclamation project is in the hands of two major international Dutch dredging and marine contractors – Boskalis and Van Oord. These two companies have created a jointly owned company called PUMA (Project Organisation for the Expansion of the Maasvlakte). In addition to using their own machines, a number of external contractors also help out and the preparation work for the new hard sea defence keeps forty-five machines – some of which have been specially developed for the project – fully occupied.

These include two modified Hitachi EX1200s. The first of these special excavators was modified by Dutch contractor Snijder; it has three main lift rams, an extra rear ballast to support the 25 m boom and stick, and a widened and lengthened undercarriage. The second modified EX1200 also needed three main lift rams to support its long-reaching front end. It came from the Dutch contractor Jac Rijk.

Snijder has also made a one-of-a-kind Cat 385 for the project. The Condor, as it is known, is fitted with a double cab and has an unbelievably long reach of 46 m. The tip of the stick is fitted with a sonic device for surveying close to the shore where it is too shallow for survey vessels and too dangerous on foot.

However, big and wonderful as these machines are, the star of the show is a 1,200 tonne giant rock handler that cost PUMA the best part of €5 million. The Blockbuster is based on an E-Crane, the creation of Belgian-based company Indusign/E-Crane Worldwide, which has built up a reputation all over the world for their dredging, scrapping and port handling abilities. The top half of the Blockbuster was sourced from a

2.21. Hover-Tracks transport the huge 40 tonne concrete blocks needed for the hard defence sea wall.

2.22. The project is the domain of a pair of highly modified Hitachi EX1200s with three main lift cylinders to carry a longer boom and stick.

2.23. This one-of-a-kind Cat 385, dubbed the Condor, features a double cab and has an unbelievably long reach of 46 m. The stick is fitted with a sonic device for surveying close to the shore.

Texan port in the US.

PUMA came up with the idea for a brand-new one-of-a-kind undercarriage. Re-using some key components – such as three of the track groups – another PUMA idea was to insert a traverse beam to connect the upper and the lower halves. In doing so, the reach of the machine was extended to the desired 50 m.

The result of these modifications not only provides a canteen area, office, workshop facilities and a separate office for the surveyor, but also a sizeable deck area to store all four attachments. The complete structure is carried on six substantial tracks.

The huge machine took roughly six months to assemble and completely dominates the skyline. Fitted with a specially developed cradle, it is not only capable

of picking up the 40 to 45 tonne blocks but also of reaching out to 50 m. At this range, it can place the blocks with an accuracy of 15 cm, which is also the reason why the machine needs 360 tonnes of counterbalance.

Despite the Blockbuster's massive dimensions, it gets its power from a mere 900 hp Caterpillar engine; however, there is no danger of running out of fuel because the massive undercarriage conceals a 20,000 litre diesel tank. The low power requirement is one of the big features of the Blockbuster because the weight of a block at the front is counter-balanced by additional weight at the rear of the crane. Connected by a mechanical link, the central section of this rear ballast alone weighs 75 tonnes, while the two sides each

2.24. The Blockbuster is the star of the show on the rock-handling front.
This unique 1,200 tonne giant cost the best part of €5 million.

2.25. The Blockbuster has the painstaking task of laying 21,000 concrete blocks,
each weighing 40 to 45 tonnes. It will take nearly a year to complete the job.

contain 18 cu m of concrete.

With more than fifteen years of E-Crane experience, operator Klaas Smits reckons this latest creation is phenomenal. 'It is fantastic and there is so much power,' he says.

It is quite a climb to the cab and the last stage takes the operator through two steep stairways and up the middle of the pylon supporting the crane. The cab is amazingly quiet, because the engine is so far away.

Offering 8 m of vertical movement at its maximum and sitting a lofty 18 m above the ground, the operator has a fantastic view over the North Sea and far along the coast. Despite its size, visibility around the machine is also good and the operator can call up images on two CCTVs from no fewer than eight cameras.

Operators have plenty of time to enjoy the views because there are always two operators, swapping roles every one or two hours. Even if they work round-the-clock with two 12-hour shifts, it will still take nearly a year to position all 21,000 blocks.

Once the Blockbuster has finished the job at Rotterdam, there is speculation surrounding what the machine will do next. At this stage it is too early to tell but, providing it is successful, then it is possible that the giant Blockbuster could be used on similar projects elsewhere.

Grensmaas

Once every fifty years the river Maas in the Dutch province of Southern Limburg bursts its banks. During Christmas 1993, disaster once again struck, causing several million euros of damage. The army was called in to evacuate 10,000 people.

In 1995, just two years later, the river was once again unable to contain its contents following another period of prolonged rainfall. This time, the evacuation of over 200,000 people made international headlines.

The rule of averages would suggest that the next flood is not due for another three decades, but with concerns about the effects of global warming, it is feared that flooding frequency may increase. Something needs to be done and quickly.

The river begins as a mere 400 m high trickle in the France Vosges. As it travels, fuelled by rainwater, it merges with waters from the Ourthe and Amblève rivers in the Belgian Ardennes. The river has to carry it all,

even during periods of prolonged heavy rainfall. By the time it reaches Limburg it flows at the rate of 250 cu m/sec. In exceptional circumstances, such as occurred during the floods of the early 1990s, this speed increased to 3,000 cu m and the water level rose by 2.5 m in just forty-eight hours.

The price for protecting the 10,000 families in the towns and villages lying close to this volatile artery is roughly €500 million. Fortunately, the Dutch have devised a plan to reduce the risk of flooding to once every 250 years, without costing the taxpayer a single penny! That may sound like a tall order considering that the creation of flood plains on a 43 km section of the river Maas between Maastricht and Roosteren, means shifting 80 million tonnes of material between now and 2023, but it is not. Gravel, not taxpayers' money, is financing the project.

Gravel is in scarce supply in the Netherlands and Limburg is the only area in the country where it can be found in a deposit laid down more than 100,000 years ago during the last ice age.

The river banks are absolutely full of gravel. In some places it piles high above the river bank. The average depth is about 8 m, but in places this gravel layer dips 15 m before hitting bedrock. It is estimated that during the past seventy years, almost 3,000 hectares have been used for extracting around 250 million tonnes of gravel. There is a downside, however, because the extraction is unpopular for leaving deep and unnatural indentations that soon fill with water.

The new fifteen-year project is different because it combines sand extraction, gravel extraction and flood protection with the development of 1,000 hectares of new nature area. The ambitious task is in the hands of the Grensmaas consortium. The Dutch word 'grens' means border and, in this case, refers to the river's role as a border. Dutch Limburg lies on one side and Belgian Limburg lies on the other. The dividing line is the deepest point of the river.

The project is being closely monitored by the Dutch government and the consortium of three partners includes gravel producers, contractors – Boskalis, Van Oord and Van den Biggelaar – and a national nature conservation organisation called Natuurmonumenten. In undertaking this project, the partners were required to earmark €4.5 million for archaeological digs, which was fortunate because the discovery of a mass grave containing well-preserved remains of nearly eighty

2.26. A pack of modified Caterpillar 385s are the prime movers at the Grensmaas flood protection and gravel extraction project in southern Holland. This one comes from Dutch contractor Snijder.

2.27. This Cat 385 wields a 22 m long boom and stick combination to which is attached a 4 cu m bucket.

horses from the sixteenth century made international headlines.

Pack of Cats

The first stage of the project is due to run until 2015 and it keeps fifty machines fully occupied. Many large Cat excavators are currently on gravel extraction duty at Itteren. Since the 3 m thick top layer of soft clay has been removed, the four Cat 385s are able to make a serious dent in digging up 11 million tonnes of gravel alongside another 2.5 million tonnes of sand.

Dutch contractor Snijder is well-known in the international earthmoving world. Operating with a rental fleet topping 300 machines, the company prides itself in modifying machines to customer requests. One of their

2.28. In the first stage of the project at Itteren, the top layer of soft clay was removed.
Now this Cat 385 – one of four – is helping to dig up around 100,000 cu m of gravel a week.

popular modified machines – a 385C – is usually on dredging duty so it has been altered in-house to have wider tracks and taller undercarriage. Fitted with a specially made 24 m long front arm and 3 cu m bucket, the 385C is able to excavate gravel from a maximum depth of 10 m.

The second 385 helping excavate at the river was supplied by Janssen Grondverzet. This machine wields a 22 m long boom and stick combination attached to a 4 cu m bucket.

Dutch contractor Jac Rijk supplied another 385, and the additional rear ballast confirms it is also heavily modified. Up front, the boom has been cut in half and lengthened. Working from 7am to 7pm from Monday to Friday, the combined efforts of all four excavators are good for 100,000 tonnes a week.

Gravel Shortage

Gravel extraction is just one part of this massive project however. Smaller excavators are in charge of loading material from the river banks, which then needs shifting to backfill the hole left when the gravel has been removed.

The extracted gravel and sand is stockpiled in a 25 m tall heap, which when full will contain 200,000 tonnes. All this material leaves from a temporary harbour measuring 250 m deep by 500 m long – a super-sized bath tub with direct canal access to Rotterdam.

Fifty-three million tonnes of gravel may sound like an awful lot but, in fact, it is only enough to meet a

2.29. A pair of Komatsu dozers helps backfill the big holes left after all the gravel has been removed. When finished, the project will provide over 50 million tonnes of gravel.

2.30. One of four Volvo prototypes developed specially for this project. Featuring an extra axle and a larger tipping body, it is capable of carrying 60 tonnes at a time.

CHAPTER 3

Wales

The Ffos-y-fran Coal Extraction and Land Reclamation Project is one of the biggest coal mining projects in South Wales. Located near the Brecon Beacons, this 400 hectare derelict site used to serve as an illegal dumping ground. Officials have calculated that the cost of removing and treating the waste from the three landfill sites would have added an extra £1,000 to the council tax bills of every household in Merthyr Tydfil.

Luckily for the inhabitants of Merthyr Tydfil, a specially created joint venture between the Miller Group Ltd and Argent Group Plc (henceforth known as Miller Argent) took on the burden of cleaning up the landfills. This was a truly mammoth task. dozens of burnt-out cars were removed and a World War II bomb was even discovered! The intentions of Miller Argent were not completely altruistic, however. The reason for their interest was the hope of uncovering approximately 11 million tonnes of high-quality Welsh steam coal in the process, a side-effect of the cleaning that would more than make up for their efforts.

Removing the rubbish was just one part of the deal, however. The other part stipulated that Miller Argent had to take out all the potentially dangerous disused iron ore and

3.1. Miller Argent has four of these powerful Komatsu PC3000-6 excavators. Equipped with 1,260 hp engines, they work twelve hours each day from Monday to Friday.

coal workings and restore the area to urban common after the completion of the mining operation. The project is a big one and has been forecast to run for another fifteen years. Millions of cubic metres of material had to be shifted to clean up the site before crews could begin to reach the coal.

Every week a dozen trains carry the fruits of hard labour – roughly 20,000 tonnes of coal – by rail from the Cwmbargoed Disposal Point to the Aberthaw power station 30 miles away on the coast at Barry. The material could potentially go by road, but shifting it by rail eliminates nearly a thousand truck journeys a week and thanks to the introduction of a brand-new multi-million pound barrel washer, this level of output is only going to increase further.

Unfortunately, the start of the operation was not a particularly smooth one. In the first year only 400,000 tonnes of coal were produced, which was not ideal considering Miller Argent's initial contribution of £90 million. There were several reasons for the slow start: geological faults and abandoned excavations proved awkward and the weather proved unpredictable. Also, progress was hampered by a shortage of on-site materials for the haul roads, which produced added difficulties because the road to the tip was a long, steep and slow climb

for the trucks. All in all, the start was fraught with problems and Miller Argent has since placed faith in the fleet of big earthmovers to make up for lost time.

Miller Argent's big earthmovers include four 260 tonne Komatsu PC3000-6s – two front shovels and two backhoes. These are the largest and most powerful excavators in Britain. It took nine vehicles to shift parts of each of these £1.5 million machines from the factory in Germany and an additional two cranes, five highly qualified engineers and loads of hard graft to put them together.

Powerful excavators, like these Komatsu machines, are essential for sites like this. The close proximity to the town of Merthyr Tydfil means they have to keep blasting to an absolute minimum; consequently, they need excavators capable of extracting the material directly from the ground. There is little straining or nodding as the metal muscles on these German-made machines force their 15 cu m buckets into the stubborn material to gouge out 30 tonnes at a time.

Four Komatsu PC3000s

The Komatsu PC3000 has an interesting history because it originally started life as a Demag H 185 in 1978; the model which was then succeeded by the S version in 1990. The

3.2. The PC3000 face shovel features a 6 m boom, 4.3 m stick and a 15 cu m bucket. The bucket stands nearly 2.5 m tall, almost 3.5 m wide and when empty weighs 23 tonnes.

3.3. The Demag H 185S is an icon of the British mining industry.
The machine pictured is believed to be one of just three active machines in the country today.

popularity of this machine with British coal mines was outstanding: fourteen of the nineteen machines sold in Europe were bought by British customers. However, years of punishing work took their toll, and only three of the fourteen remain in operation today. Miller Argent has one of them.

The H 185S went on to become the H 255S in 1998, before hatching into the PC3000-1. The modern-day PC3000-6 model has been a successful machine for Komatsu, especially in the UK. Of the 178 machines

3.4. The PC3000 is a successful machine for Komatsu. Four of them work at Ffos-y-fran.

3.5. *The Cat 824H wheeled dozer is not a common sight in Britain but this 400 hp machine is the preferred choice for cleaning up around the big excavators.*

produced in Germany so far, at least two dozen can be found in Britain, but the Ffos-y-fran site no longer boasts the only ones in Wales. (See Postscript.)

However, the four PC3000s are not the only big excavators on site. Miller Argent recently bought a Cat 385, an 85 tonne machine, to help a Komatsu PC1250 clear the last layer of overburden from the coal. Smaller than the PC3000, this 115 tonner is useful for cleaning out difficult corners and thinner layers of inter-burden. In fact, it is preferred to the larger, heavier excavators because the big machines are more likely to force some of the inter-burden into the seam.

Noise and Dust Reduction

The Ffos-y-fran Coal Extraction and Land Reclamation Project provides up to 600 direct and indirect jobs, and will ultimately pay out almost £70 million in wages, but the proximity to Merthyr Tydfil is an issue. Some local residents have complained about the noise of the big machines. Complaints like this are not unique to this site. All European mining operations are under pressure to reduce the impact on their surroundings. Miller Argent takes these noise issues very seriously and is spending a fortune on ways to make their machines quieter. The Komatsu mining excavators are already available with factory-fitted noise reduction kits which were originally developed several years ago for work in Australia, where noise issues have always been a lot more sensitive. Unfortunately, it is too difficult to retro-fit these factory-fitted systems to machines already out in the field.

Something different was needed for the excavators owned by Miller Argent and a number of concepts were evaluated, one of which includes a new cooling-radiator fan-assembly. Developed in conjunction with KMG Warrington, the UK distributor for Komatsu Mining Germany, the new polypropylene fan design is lighter and more efficient. Although it still pumps the same volume of air, it does so at a slower speed, which helps to reduce noise levels. The design is now fitted to all four excavators. One of their two backhoe excavators also features a new quieter running fan design and a couple of other novel noise-suppression devices.

3.6. Like all mines and quarries, Miller Argent is constantly exploring new ways to reduce the noise of their big machines. This PC3000 features several improvised noise reduction kits on the upper structure.

Wherever air is pumped there will be noise. One of the noise-suppression concepts successfully tested is the attenuation box, which can be bolted to the sides of the body of the hydraulic excavators and is designed to reduce noise as air is sucked into the engine radiator. Inside these attenuation boxes, sound waves bounce back and forth until they lose their energy. Reducing engine and radiator coolant fan noise are two areas where improvements can be made but the oil coolers are the biggest source of noise on any hydraulic excavator. Early results seem promising so it may be that all excavators on the site will soon be fitted with attenuation boxes.

Another major noise issue that Miller Argent has had to contend with is the noise of the large engines in their fleet of trucks. When they struck a record-making £65 million deal with UK Cat dealer Finning, Miller Argent

bought forty-three new 1,000 hp trucks. Understandably, when working flat-out to haul materials to the tip, these big trucks can be a bit noisy so Miller Argent has experimented with a number of ideas to reduce this, including blanketing the engine. Unfortunately, the effects were minimal so they attempted to re-route the Cats' exhausts through their tipping body, which proved more successful. The re-routing method has since been incorporated in all Cat 777s on the site.

Noise management is not the only issue facing Miller Argent in their quest to attain the high-quality Welsh steam coal; they are also faced with the problem of dust. Joint managing director James Poyner says their dust-suppression units take care of the haul roads but the bigger problem is caused when excavator buckets empty their contents into the trucks. After careful consideration

3.7. The Ffos-y-fran project operates the largest single fleet of new Cat trucks in Europe.
There are forty-three in total – fifteen 775Fs and twenty-eight 777Fs.

on the part of Miller Argent, they eventually opted for an Italian-made misting cannon. Adopting the technology used in snow-making machines, the fan, which is 1 m in diameter, blows tiny droplets of atomised water though an arc of 325 degrees and to a distance of 60 m. The lorry-mounted, self-contained prototype also houses a water tank. James is convinced it is the solution. 'The water droplets gather dust,' he says. 'This makes them heavier so they fall back to the ground.'

They were so pleased with the results that they made a second system. A major feature of this newer machine is the massive working area which allows the water truck to park further from the excavator, thereby helping to improve safety. Working at full power the truck drains the water in just thirty-five to forty minutes, at which point, it has to stop and travel to a refilling point.

One interesting point about the new water truck is that the hoses do not need to be permanently engaged – something that operators are still experimenting with. For instance, nozzles do not need to be switched on when a truck is reversing into position, which suggests that it may be possible to switch them on and off intermittently, helping to prolong times between

refilling. Both systems can be activated remotely from up to 50 m away. Management considered getting excavator operators to control the fans but this idea was dismissed because it was yet another task for operators to have to handle.

Despite a few minor teething problems, they are so pleased with the second misting cannon that they are now toying with upgrading the original machine and swapping the water tank for a larger one. Unless the weather forecast is torrential rain, the two cannons are in use nearly every day. Even during the winter months operators reckon that the cannons are used nine days out of ten.

However, dust suppression comes at a price. Each misting cannon cost the company between €30,000 and €50,000, prices that do not include the water tanks, trucks and hours of modification work needed to marry them.

The purchase price is just one part of the entire cost because they also need specially trained operators – two per shift. When added to the rest of the seven-strong dust-suppression fleet, each day the water team needs nine operators! Dust suppression is a costly business. However, they have no doubt it is a price worth paying.

*3.8. This Welsh mining operation has employed water cannons –
generally used to make artificial snow at ski resorts – to keep on top of dust.*

No matter how dry the conditions, their excavators can keep working in the knowledge that dust is not a problem – they are on top of it.

There is also one other benefit of the investment. Now that dust is under control at the excavators, the rest of the dust-suppression fleet can concentrate on the haul roads, helping to take some of the stress out of the operation.

Finally, it does not take a mathematician to work out that two misting cannons is not enough for their four excavators. The problem could be easily solved by buying more; something that management does not rule out as the site continues to expand.

3.9. The misting cannons help to subdue dust when the excavator loads a truck.

3.10. There are estimated to be only a dozen PC1250s left in Britain and all are SP types, meaning they have a short boom and stick with a big 6.7 cu m bucket.

Hitachi EX1900

Many of the collieries and pits in South Wales closed during the second half of the twentieth century but the story of Welsh coal is far from over. As oil and gas prices continue to rise and renewable power sources are still unable to quench our insatiable energy demand, coal has once again become fashionable. Fortunately, South Wales has plenty of reserves, and the 1.5 million tonnes of coal produced from Welsh surface mines each year generates nearly fourteen per cent of the UK's energy requirements. The Miller Argent site provides 0.5 million tonnes of this, while Celtic Energy supplies an even larger volume of around 1 million tonnes. Almost sixty per cent of that goes to the Aberthaw power station and the rest is sold to the domestic and industrial markets.

Celtic Energy was founded by former British Coal directors and the company owns the mining rights to three active coal mines in South Wales. One of these is

Selar, a big operation especially by Welsh standards. Selar is located in the Neath Valley, not far from the town of Merthyr Tydfil. On a good day, it offers stunning views of the Brecon Beacons; on a bad one, visibility can be zero and it can be very wet.

Only a couple of years ago, twenty-six 90 tonne capacity Cat 777s could also be found at the Selar site. Today, just a couple remain because the mine is winding down; however, Celtic Energy is seeking planning permission to extend the site. If successful, Selar North could yield a million tonnes of coal during the next four years. If not, then the current void, which has produced over 4 million tonnes of coal during the past nine years will soon need to be back-filled with nearly 18 million tonnes of material, a task that will take three years to complete.

Until recently Selar used to be the home of an

3.11. The delivery of this EX1900-5 was a major triumph for Hitachi. It is the company's largest hydraulic excavator in the UK.

3.12. The EX1900-5 was made in Japan and launched at the Bauma show in 2004.

excavator that made British earthmoving headlines for being the very first of its kind in the country; the machine is the Hitachi EX1900-5. The 190 tonne Dash 5 was purchased as a replacement for a Demag H 185S and supplied by UK distributor HM Plant. Over the past few years the excavator stayed out of the spotlight and did some serious digging, and it seems to have successfully impressed everyone. Engineering manager Mark Heames described it as a fantastic machine. 'It didn't stop and never broke down,' he says. 'It was very reliable and we were pleased with it.'

There is one noticeable design flaw with the Dash 5 and the damage to the sheet metal at the rear displays this. Apparently, the damage is the result of rocks riding up onto the tracks during travel; if the operator fails to spot them then they grate along the bottom of the upper structure and cause the damage. They reckon this is due to insufficient clearance. 'It is purely cosmetic but it is not good,' says Mark.

Warranty played a major role in the decision to purchase Britain's very first seriously large Hitachi excavator. Celtic Energy managed to negotiate a 10,000-hour package, which is roughly equivalent to three years' operation. This was a great deal because the mine only incurred servicing costs initially.

After 10,000 hours of punishing operation at Selar, Celtic Energy moved the Dash 5 out to Nant Helen, where it clocked up another 2,000 hours before being sold. The full details on the replacement may never be disclosed but the simple fact is that the substitute does not come from the Hitachi line; instead the mine has opted to buy the very first 200 tonne Komatsu PC2000 in the country. (See Chapter 5 for more information about this excavator.)

First Impressions

As well as Selar, Celtic Energy owns and operates two other large Welsh coal mines. One of these is Nant Helen and, as with Selar, the company is seeking planning permission to extend the site.

The third coal mine owned by Celtic Energy was originally called East Pit, a former British Coal deep

3.13. This Komatsu PC3000-1 works at East Pit East Revised.
The powerful 250 tonne machine is coupled to a 15 cu m bucket that fills Cat 777 trucks in three to four passes.

3.14. MST uses only the highest quality Swedish steel to make its buckets.
The 12 cu m bucket pictured here should provide up to 12,000 hours of trouble-free operation.

mine. The site is now called East Pit East Revised and it is located in the upper Amman Valley at the northern end of the South Wales coalfield. The mine is very much in its infancy, containing seventeen coal seams that are estimated to yield around 2 million tonnes of coal, which will need to be excavated in seven years.

Unfortunately, Celtic Energy was faced with a big problem in the early stages of the newly purchased mine. This problem was water – the bottom of the mine was full of it! In places it was more than twenty metres deep. The mine's trucks needed to haul overburden to the bottom of the void but they were unable to do this because the void was completely full of water caused by rainfall and inflows from deep-mine workings. In the end, Celtic Energy had to have huge pumps working round the clock to drain the artificial lake.

East Pit East Revised is the home of a number of big earthmovers. One of these earthmovers is a Komatsu PC3000-1, an older excavator that replaced the Demag H 255S when Komatsu Mining Germany completed the purchase of Demag in 1999.

Not all the machines working in East Pit East Revised are old ones. Another of their Komatsu PC1250-8s is still in full employment. Although the excavator was made in Japan, the bucket was supplied by British

company Midland Steel Traders (MST). This Sunderland-based company claims to make the best mining buckets available. Time scales vary depending on bucket size, but guidelines state that it takes around three months to make a single bucket. Discussions and drafting drawings can take a month and another month is needed for preparation work. Then, using the strongest Swedish steels, it can take four weeks to construct the bucket. The company gives its buckets a 6,000-hour or one-year warranty – whichever comes first. They claim, however, that most will stay on a machine for at least 12,000 hours before needing any repairs. A bucket's major strength or weakness is where it attaches to the excavator; as long as this stays intact there is no reason why it cannot survive for 18,000 hours.

Elsewhere in East Pit East Revised, another £1.5 million has been spent on yet another Hitachi EX1900 excavator, the Dash 6. This Dash 6 version was bought as part of a plan to double current coal volumes to 7,000 tonnes a week. This is a major challenge, because many tonnes of overburden need to be stripped in order to expose just a single tonne of coal. That ratio will drop in the future, but for now, the three big machines and seventeen trucks need to be shifting around 140,000 tonnes of material a week.

3.15. The second of Celtic Energy's EX1900s works flat out at East Pit East Revised – a former deep British Coal mine. Pumping away at the heart of this excavator is a 38-litre Tier 2 Cummins engine that generates a maximum of 1,086 hp.

The new excavator is not too dissimilar to the earlier Dash 5 model. Even an untrained eye can spot the grey stripe around the bottom of the upper structure. Operator Paul Evans is proud of his machine, though he has had to adapt to a number of differences between the Dash 5 and Dash 6. The big difference for operators like Paul is in changes to the excavator's cab. The Dash 6 is wider, leaving more space for the operator. There are new instrument panels and many more computers on this machine. 'It is far more sophisticated,' says Paul, who is not entirely convinced this is the best way forward. 'If I am honest, I prefer the simplicity of the Dash 5 excavator.'

One thing that does impress Paul is the average fuel consumption figures of the Dash 6, which suggest that it consumes less than 140 litres per hour. 'This is only marginally higher than the Dash 5 machine and given that the Dash 6 has a bigger engine and has to contend with tougher rock, this is not bad.' He is also impressed by the addition of smaller, electrically operated wrist-type joysticks. 'They are much lighter and less tiring to operate.' Finally, the fact that the Hitachi Dash 6 can be coupled to the any of three large MST buckets (8, 9.6 or 12 cu m), while the Dash 5 excavator was only offered with a 12 cu m bucket, makes the newer model a hit on the site.

However, Paul does highlight some negatives to operating the Dash 6. He says he misses that little bit of extra finesse needed to combat the resistance of the Dash 5 servo-control joysticks. 'It is sometimes difficult

3.16. East Pit's EX1900-6 is opening up a new dig and there is not a lot of room for the trucks to manoeuvre. Once the excavator is properly set up it will shift 650 cu m of material an hour.

to get precise control on this one; it takes some getting used to.' Also, he is not keen on the fluffy carpet arrangement covering the cab floor of the Dash 6. 'It is always full of dust and really difficult to clean.'

Currently notching up sixteen hours at the front line every day, by the time the excavator hits the agreed-upon 10,000-hour warranty mark Celtic should be midway through the mine's life. It remains to be seen what brand of excavator it will be swapped for to complete the job. And what a big job! Even when coal reserves have been exhausted, it will still take an additional four to five years to backfill the void with 80 million tonnes of material.

3.17. Operator Paul Evans enjoys the new joysticks on the EX1900-6. He thinks that the Dash 6 model is much lighter to operate than the older Dash 5.

CHAPTER 4

Finland

A blanket of forest covers almost seventy per cent of Finland and, until recently, the main industry in this area was the production of paper. Originally, paper mills provided important jobs for many isolated and rural communities but over the past few decades an increased reliance on machines within the industry has led to a rapid decline in the need for workers. To make matters worse, neighbouring Russia has muscled in on the industry by providing cheaper pulp, which, when coupled with the fact that several paper mills in the area have already been forced to close, means that the future of the Finnish paper industry is seriously under threat.

Fortunately, the country's surface mining sector is making a recovery and many redundant forestry workers have found work in a number of newly created operations. One of these is in Finnish Lapland, which covers 100,000 sq km, or a third of Finland's land mass, making it the least populated area of the country. Sixty thousand of the 200,000 inhabitants of this isolated area live in Rovaniemi – the capital of Finnish Lapland. Winter tourism is a valuable source of income for the people of this area but some people still manage to make a living from gold panning.

This gold panning is done on a much larger scale at the biggest of the three Finnish mines, Agnico-Eagle. Officially opened in June 2008, this gold mine lies north of the Arctic Circle, close to the small town of Kittilä. The first deposit of the highly prized substance was discovered in 1986, but it was not until 2004 that Canadian company Agnico-Eagle Ltd bought the rights to mine it.

Providing nearly one hundred badly needed jobs, the first 25 kg gold bar left the site in January 2009. The gold is not exactly visible at this mine because every tonne of ore contains just 5 g of gold; this means that in order to mine the 5 tonnes of gold that they seek each year the men and machines need to dig up a million tonnes of ore. Fortunately, with ore reserves in excess of 18 million tonnes, there is plenty there for the taking. During the

4.1. The Agnico-Eagle gold mine is situated above the Arctic Circle in Finnish Lapland in one of the most remote areas in Northern Europe.

4.2. Three Cat 385 excavators will extract roughly 10 million tonnes of material a year, which yields about five tonnes of gold. During the next fifteen years they will produce more than 90 tonnes of the precious metal.

next fifteen years the mine confidently expects to produce more than 90 tonnes of gold and though all of this cannot come from the six open pits alone, there are no plans to go deeper just yet. Currently the plan is to concentrate on the two surface mines for the first two years, the largest of which is estimated to yield a total of 3.5 million tonnes of gold ore.

This precious material is being extracted from the Agnico-Eagle mine by a brand-new fleet of Caterpillar machines. These include three 85 tonne 385C excavators which need to fill seven 777F trucks with 3 million tonnes a year. The excavators may appear small for the job, but there are good reasons for the choice. One reason has to do with the fact that the mine's benches contain mixed levels of ore and waste so the small 5.6 cu m buckets are ideal for the job because they allow the operator to pick and choose what they excavate. More importantly, the stripping ratios of this particular mine are forecast to drop dramatically after a couple of years, as the mine prepares to move underground, and when this happens the big machines will no longer be needed. In practical terms, three 85 tonne excavators should be much easier to sell than a couple of big 200- or 300-tonners.

In fact, it is not the gold or the Cat excavators that are attracting all the attention at Agnico-Eagle; instead, a cleverly designed transport system is stealing the limelight. One thing that all large tracked excavators have in common is that they are not easy to move. That these machines can travel at speeds of just 1.5 to 2 km/hr is the main reason why most mines opt to carry out their servicing and repair duties in the pit. However, the Finnish have cunningly devised a new system whereby these machines can be moved without additional wear and tear to their undercarriages.

The new system was invented by Ossi Kortesalmi, who used to operate a wheeled Broyt excavator for contractor Mäntylä Mining at a large chrome mine. In Finland it is quite common for quarries, mines and even construction sites to use an ADT or rigid dump truck to move wheeled excavators. The operator has to force the bucket down into a heap of material inside the tipping body of the truck, then the excavator's front wheels are raised clear of the ground and the truck can drag the excavator around the site. This was Kortesalmi's experience and he pondered utilising a similar method to speed up the transport of the mine's three tracked excavators – a Cat 375, Hitachi 800 and Kobelco SK1340.

4.3. Three Caterpillar 385s do all the main digging at the mine. Each one must shift nearly 3 million tonnes of material a year.

After drawing a number of preliminary sketches, Kortesalmi enlisted the help of a local engineer in creating a prototype. By early 1996, it was ready for testing with the two 85 tonne excavators. In Kortesalmi's design, the original idea of placing the excavator bucket into the dump truck bed was kept and an additional solution was added to protect the rear of the machine. He added a pair of rear axles and ramps onto which the excavator tracks rested. The design worked so well with the Cat and Hitachi excavators, that a year later he made a heavy-duty 140 tonne version for the Kobelco excavator.

Encouraged by these early successes Kortesalmi established a company called Sleipner Oy, which was located at Tornio, the border town dividing Finnish and Swedish Lapland. After a number of years spent refining his design and obtaining global patents, he unveiled it at the Bauma show in 2004. This show put his invention, which featured the addition of safety brakes, under the international spotlight for the very first time.

Kortesalmi's design was dubbed the Sleipner Transport System; Sleipner being the eight-legged horse of Odin in Norse mythology. Over the years, he continued to refine the Sleipner Transport System, even creating versions for

70, 90, 120, 190 and 250 tonne excavators, and the results were unveiled at Bauma in 2007.

The system continues to attract attention from major mines and there are currently more than thirty systems at work. Spanning in weight from 70 to 250 tonnes, they have been bought by customers in Germany, Ghana, Israel, Indonesia, South America, South Africa and even the United Arab Emirates.

Agnico-Eagle mine superintendent Juha Riikonen confirms that the system provides a quick way to transport the excavators to the workshop at speeds of up to 20 km/hr. It is used almost daily, so he is confident of a quick return on the €65,000 investment. 'I am surprised by how little the concept has caught on worldwide,' he says. 'Compared with the value of the machines that it moves, the Sleipner Transport System is relatively inexpensive.'

The way that the Sleipner Transport System protects the back of the excavators is by fitting each axle and wheel arrangement with two tyres. Although thirty per cent of the excavator's weight rests on the truck, each of the two wheels still needs to support over 30 tonnes. Normal truck tyres are not strong enough so port-handling tyres are used. The set-up then works as follows: with the rear of

4.4. It is not every day that you see an 85 tonne excavator hitching a lift on the back of a dump truck. The Sleipner System, employed by the Agnico-Eagle gold mine, greatly reduces journey times to and from the workshop.

the excavator's tracks parked on the axles, the excavator's upper structure turns to face the truck, which must be twenty to thirty per cent loaded. The excavator then depresses its bucket into the heap of material in the bed of the dump truck and the front of the tracks are raised clear of the ground. The material in the truck functions as a sort of bearing: placing the bucket directly onto the bottom of the tipping body would not work because in addition to abrasion the bucket could slide forward when a truck travels downhill.

Arctic winters are tough and temperatures regularly plummet to minus twenty-five degrees Celsius, but if the wind blows in from Siberia then it can drop as low as minus forty-five degrees Celsius – the lowest temperature ever recorded at Rovaniemi. The mine's haul roads, which are maintained at a gradient of ten degrees, can be covered by a layer of snow or ice from November until late April. The frequently asked question is how do the trucks brake when going down a slippery haul road with an 85 tonne excavator? The mine authorities confirm that this is not a problem for their operators because they can lower the front of the excavator's tracks to provide extra braking performance when necessary. The Agnico-Eagle mine is definitely convinced of the benefits; they always use the

Sleipner Transport System to bring the excavators to the workshop and, when blasting, it provides a quick and easy way to shift the machines to safety.

Sleipner Oy has now turned its attention to developing their system for even larger excavators. Most recently they created one for 400 tonne excavators and a number of units have already been sold, including one to a mine in Bolivia with a Komatsu PC4000. Although similar in design to the smaller systems, these heavy-duty Sleipner systems use beefed-up components and are fitted with eight wheels. It is not only the weight of each axle unit that has been increased (20 tonnes), but also the price to over €300,000 a pair.

The company sees the biggest future potential in developing systems for 250 to 400 tonne excavators. 'Undercarriage maintenance costs on these big machines are high,' says Ossi, 'especially when they are expected to move around frequently and for considerable distances.' He is also looking to develop an even bigger version for 550 tonne excavators. This really will be a sturdy device because 385 tonnes of the excavator's weight will need to rest on the transport wheels. It is quite likely that this next version will need eight 27.00.39 tyres – the same size fitted to a Cat 777 truck!

Phosphate Mining

Every country has an earthmoving contractor that is just that bit bigger than the rest. In Finland that contractor is the family-owned Hartikainen business which operates a 200-strong machine fleet with some hefty excavators. A mine in dire need of massive excavators is the Yara-owned Siilinjärvi phosphate mine, located just north of the attractive city of Kuopio, an hour's flight or a 450 km drive north of Helsinki. There are larger phosphate mines in Russia, but this Finnish one is estimated to be the largest in Europe. Currently measuring almost 3 km long and 700 m wide, at its deepest point the Siilinjärvi mine is 190 m below the surface. Drill holes have revealed phosphate ore to a depth of 800 m and the deposit extends for more than 16 km. Main contractor Hartikainen has been active here since the very start of the mining operation in 1979.

There should be reserves at this mine for at least another thirty years but the phosphate is not easy to obtain. The current pit is hemmed in by a railway to the south and a reservoir to the north. Therefore, lengthening the pit will be difficult, so future plans are to increase the width by over 1 km and the depth to 300 m. It is also likely that other smaller pits will need to be opened and that the whole operation will shift underground in the future. However, for now there is plenty of phosphate ore above ground.

The 10 million tonnes of phosphate ore extracted each year are conveyed to the production facility on the other side of the road where 850,000 tonnes of phosphate concentrate is converted into 300,000 tons of phosphoric acid. However, Yara wants to increase this figure by another 150,000 tonnes of phosphate concentrate a year; in order to do this they need to produce a million tonnes of phosphate concentrate, which means shifting 18 million

4.5. The Siilinjärvi mine is believed to be the largest phosphate mine in Europe. Future plans for expansion mean the big machines will need to be shifting 18 million tonnes of material a year.

4.6. The older EX1900-5 is on overburden extraction duty. This is the machine that was on Hitachi's stand at the 2004 Bauma show.

4.7. Hartikainen's Hitachi EX1900-6 is temporarily out of action for a service.
Easily scooping up 20 tonnes at a time, it can fill a Cat 777 in five passes.

4.8. *The 12 cu m bucket fitted to Hartikainen's EX1900-6 was made in England by MST.*

more tonnes of material. Ultimately, their goal is to increase phosphate ore production to 15 million tonnes per year.

With so much material to shift, Hartikainen have always needed plenty of forceful diggers. They were the first Finnish company to own an excavator with an operating weight of more than a hundred tonnes. The excavator they chose was a brand-new Hitachi EX1100 which was delivered to the mine in January 1988. Just one year later it was joined by a second one, and together these two machines did most of the work at the Siilinjärvi mine for the next five years. Today, the oldest excavator has notched up nearly 50,000 hours, while the other machine has provided over 40,000 hours. However, even the combined efforts of these two powerful machines were not enough to keep pace with increased production forecasts and it was decided that additional digging force was needed.

Hitachi was the obvious choice because Hartikainen was pleased with the long and reliable service life of the two EX1100s. 'We know Hitachi machines and our operators like them,' says Arto Hartikainen. 'So it was an easy choice to stick with them.'

In the end, he opted for an even larger 190 tonne Hitachi EX1900, which was delivered in July 2004. This Dash 5 version was plucked straight from Hitachi's stand at the Bauma show. During the next few years, they also added two new 110 tonne EX1200-5s – the first in February 2002 and the second in February 2008 – to the fleet. A 100 tonne Cat 992G wheeled loader was the next addition to the growing fleet; these machines provide a much quicker alternative for loading ore when rapid changes of position are needed. Finally, in April 2006, Hartikainen bought yet another brand-new EX1900; this Dash 6 version effectively replaced the two older EX1100s.

The phosphate ore operation works 24/7, while the waste extraction teams operate two shifts per day from Monday to Friday. In total there are fifteen heavy-duty dump trucks – thirteen Cat 777s and two 91 tonne Komatsu HD785s – at Siilinjärvi. The new Dash 6 model excavator will be expected to perform at least 7,000 hours of operation a year for Hartikainen. The owner will be unhappy with anything less than 30,000 hours during the next five to seven years. 'Preferably 40,000 hours,' Arto says.

As for the fate of the two older EX1100s, it is still to be decided. 'We thought about cannibalising one to keep the other going,' he says, 'but we will probably just sell them both.'

4.9. A Cat 992G provides the mine with additional loading flexibility. Phosphate ore is abrasive and tough on tyres so chains help to increase their life expectancy.

4.10. This machine is one of two EX1200s at the phosphate mine. The 6.5 cu m bucket fills 100 tonne trucks in eight or nine passes.

Winter Hill Wonders

The two EX1900s at the Siilinjärvi site are big, but they are not the largest Hitachi excavators in Finland. The largest excavators work at another brand-new site called Talvivaara (Winter Hill), a nickel and zinc mine that will soon provide work for 500 full-time employees. Located at Sotkamo, to the east of Kajaani, Talvivaara is just 100 km from the Russian border. It is a remote location and one that the inhabitants share with lynx, brown bears and even grey wolves.

The operation is owned by the Talvivaara Mining Company Ltd (TMC) which purchased a sizeable chunk of 61 sq km of land. Originally a greenfield site, it is astonishing just how much progress has been made at this mine in the three years that it has been operational. During the first two years, Talvivaara kept nearly 2,000 contractors and staff and a staggering 350 machines fully employed. Locals used to joke that the second largest industrial project in Finland was a black hole: it sucked everything in and nothing came out!

A large chunk of the initial €450 million start-up cost was invested in the support machinery and buildings, such as the processing plant, crushers, conveyors and the leaching plant. Mine management breathed a huge sigh of relief when the first 100 tonnes of nickel sulphide concentrate left the premises. Now the mine operates at full speed due to the fact that mine owners have negotiated a ten-year contract to sell everything they can extract, much of which will be used to make stainless steel. Still, the start-up costs are strikingly insignificant when you consider that the mine predicts an annual turnover of €350 million. Nickel production accounts for seventy-five per cent of annual income, although Talvivaara will also produce a percentage of zinc.

Much of the forest in the immediate area around Talvivaara has already been cleared and contractors' machines continue to expose the ore. Although much of it can be found just under the surface, Destia – another large Finnish earthmoving contractor – has a four-year contract to pull out 2 million tonnes of overburden a year.

Mine manager Arto Suokas revealed that the output of 15.3 million tonnes of nickel ore and 7 million tonnes of waste in 2009 made it the biggest mine in the country. The ore is of a low grade and contains just 0.25 per cent nickel sulphide, meaning that each tonne excavated contains just 2.5 kg of nickel. This may not sound like much but TMC is confident of eventually extracting eighty-five to ninety per cent of the ore reserve, which is estimated at 1 billion tonnes. Also in their favour is that they have fifty years to complete the job and nickel output is expected to rise quickly during the next couple of years. 'We need to shift 25 million tonnes of overburden a year and a further 21 million tonnes of ore,' says Arto. 'At peak production we hope to handle almost 50 to 60 million tonnes of material a year!'

Currently the world's largest nickel mines are located in Russia, Canada and Australia but when Talvivaara production hits 21 million tonnes of ore, their nickel output will rise to

4.11. Opening mines provide new jobs that are vital to the Finnish economy. The Talvivaara mine will need to employ 500 people to operate at full capacity.

4.12. Talvivaara has plenty of material to blast. Once the entire excavation project is complete the hole will be approximately 5 km long and 1.5 km wide.

45,000 to 50,000 tonnes. Equivalent to five per cent of global production, this will make Talvivaara in Finland the world's largest single nickel surface mine. Their fleet of mega movers are currently just scratching at the surface but eventually it is planned that the mine will drop to a depth of 600 m. It may take a couple of decades but it will grow to a length of at least 5 km and will be 1.5 km at its widest point. 'It may get even bigger,' says Arto. 'We are still not fully certain of the extent of the ore body.'

The ore at Talvivaara is not only abrasive, but also quite slippery due to high graphite content. The only solution to this problem is to blast and ensure that the material is as fine as possible. During the start-up process, the mine was plagued with crushing problems as a result of this slippery rock. When machines were trying to break up blocks, larger blocks that would normally pass straight through the gyratory crusher tended to bridge. It was a frustrating time for all concerned. Although the crusher has the capacity to crush 3,500 tonnes per hour, during the first year it struggled to do half this.

With that problem resolved the big excavators are working flat out to meet the insatiable appetite of the gyratory crusher and, weighing 360 tonnes per machine,

these excavators are massive. TMC put some of the initial funds toward buying the only two Hitachi EX3600 excavators in Scandinavia: together they cost around €8 million. Their sole role is to clear up the debris from the biggest blasts. TMC has the Dash 6 version of this excavator, which was launched in 2007. There are a few differences between this new model and the Dash 5, the most visible concerns the new livery and the grey stripe. Inside there have been a few changes too, the Dash 6 machine has replaced the Dash 5 Mitsubishi engine for a 1,944 hp Cummins QSK 60 and its servo-assisted hydraulic system for an all-new electro-hydraulic version. Main benefits of the Dash 6 include less in-cab noise, faster response and less operator fatigue. The excavator was born out of the EX3500, of which Hitachi made just over a hundred units. The Dash 5 unit in particular totalled around eighty-six units and Hitachi's high expectations for the giant digger appear well-founded because well over a hundred Dash 6s have been delivered so far. They aim to produce approximately forty of the Japanese excavators a year. Each excavator is capable of loading 3,000 tonnes per hour, which is just as well because each year a single machine needs to shift

4.13. At the Talvivaara nickel mine, the two Hitachi EX3600s are the star attractions.
They were the first EX3600s in Scandinavia and have an operating weight of 360 tonnes.

4.14. The 20.65 cu m bucket of the EX3600 was made in Britain by MST.
Even empty it weighs 22 tonnes and can lift 40 tonnes at a time.

4.15. The two EX3600s are each expected to load 12 million tonnes of material a year each.

around 12 million tonnes of material and they typically have 18.5 production hours a day to do this. With a density of 2 tonnes per cu m the excavators each extract nearly 40 tonnes at a time. The 20.65 cu m bucket is well up to the job but the abrasive nature of the material means bucket wear is high. The material is so tough that it soon gouges out the mild steel welds, meaning that one of the three buckets is always in the workshop.

The EX3600s were bought with a five-year service contract, and Finnish Hitachi importer Rotator has invested heavily to ensure their machines meet the targets. As well as installing nineteen staff at the mine, a figure which includes fourteen mechanics, they hold an impressive €500,000 stock of spare parts.

'It is a big commitment for us,' says Rotator top man Thomas Åhman. 'Maintaining the Talvivaara machines is by far the biggest service project for us in Finland.' The mine is looking for at least 48,000 hours of operation from each of the giant excavators within the next seven to eight years. They are keen to find out whether Rotator's machines will meet these demands.

Mountain Mover

The Talvivaara mine may have two of the biggest excavators in Scandinavia but that does not mean they are not interested in acquiring more big earthmovers. Dubbed the Mountain Mover, Komatsu's WA1200-3 is no ordinary

wheeled loader; it is the world's largest mechanical one, with an operating weight of 205 tonnes and a price tag of around €3 million. This massive machine is the result of a long period of development which started when Komatsu purchased Dresser's portion of the joint venture back in 1994. At the same time Komatsu ceased production of the Haulpak 4000, an 18.3 cu m wheeled loader, in preparation for the new endeavour. By the middle of the same year, design work started on the new 20 cu m loader – a process that continued for the next two years.

In December 1997 the first prototype was built and tested in Japan. A year later the company was confident enough for a Japanese customer to try out a second prototype, and after further US tests, the machine was ready to go into production in 2000. Komatsu aims to make one machine a month at their Ibaraki facility in Japan; this same factory also makes the smaller WA700, WA800 and WA900 wheeled loaders, as well as 60 tonne dump trucks. So far the largest populations of WA1200s can be found in North and South America and Australia. In the last decade Komatsu has sold over ninety WA1200s and the machine with the serial number seventy-nine was the very first to work in Europe.

The WA1200s share the same basic engine as the Hitachi EX3600 excavator, which, in the wheeled loader, works at a top power of 1,715 hp at 1,900 rpm. Loosened material offers little resistance to the 20 cu m bucket, which even

4.16. The WA1200-3 is the largest mechanical-drive wheeled loader in the world at a length of 18 m and an operating weight of 205 tonnes. The 20 cu m bucket weighs nearly 24 tonnes even when empty.

4.17. The ground trembles when the massive 20 cu m bucket forces its way into the material. It regularly scoops up more than 40 tonnes at a time.

4.18. The Bridgestone tyres fitted to the Komatsu WA1200-3 stand almost 4 m tall. Each one is protected by a 4 tonne chain.

when empty weighs almost 24 tonnes. When specified with the high-lift option this wheeled loader is well matched to filling the world's largest dump trucks: with the standard boom raised to full height it stands a little under 12 m tall. The bucket has a rated capacity of 36 tonnes, though the mine confirms that it regularly loads 40 to 45 tonnes at a time and has no problem filling their big Hitachi trucks in just four passes.

Fuel consumption is a good indication of the performance of a big earthmover and the WA1200 burns an average of just 160 to 180 litres of diesel per hour, working in relatively light conditions. Seppo Kankaanpää of Finnish Komatsu importer Suomen Rakennuskone Oy says it is not yet working at full capacity. 'We will not be happy until it hits 250 litres per hour,' he says. 'Then we will know it is really working hard!'

Although the mine has still not challenged its new wheeled loader with a difficult job, it is clearly a machine with a ferocious appetite. Komatsu figures reveal outputs as high as 3,500 to 4,000 tonnes per hour in Australia, which is promising because Talvivaara expects it to shift 12 million tonnes of material a year.

A giant loader needs giant tyres to match. The WA1200 took a small compressor almost a day to inflate each of its 4 m tall tyres to 8 bar. Big tyres are much easier to source and a lot cheaper than they were a few years ago but they still cost a fortune. Hence, the WA1200 wears 4-tonne

chains, which bolster the operating weight to over 220 tonnes, and should ensure that the tyres last for 15,000 operational hours. The only downside to these steel protection devices is that they restrict maximum forward speed to 15 km/hr whereas without chains the loader would be able to run at 20 km/hr.

The WA1200 has lots of features one would expect for a loader of this size – such as three cameras and an extra feature which was added because the 5,000 litre fuel tank is built into the rear counterweight. An engine fire could spell disaster. To remedy this, both the engine and hydraulic chambers have been separated by a fire plate, and a local Finnish supplier has fitted automatic fire extinguishers in both compartments.

The mine's location just south of the Arctic Circle means it escapes the worst of the winter weather further north; however, temperatures regularly plummet well below zero so TMC opted for the electric heating system in their WA1200. When hooked up to a power supply this uses a series of heating elements to ensure that the engine, transmission and hydraulic oil do not freeze. Six staff have been trained to operate the fierce machine. Operator Jaakko Karjalainen is one of them. He used to drive smaller Volvo and Cat wheeled loaders and enjoys the size and power of the new machine. 'It offers great mobility and phenomenal power,' he says.

Another operator, Taisto Kettunen, agrees that he is also

4.19. Operators think that the joystick steering system on the WA1200 is less tiring and easier to operate than a conventional steering wheel.

impressed by the gentle giant. Although he used to operate a Cat 988F, the powerful Komatsu version is a big step up the power and size scale and a completely new experience for him. He particularly likes the joystick steering. 'It is great,' he says. 'Joystick steering is easier and much less tiring than operating a steering wheel.'

Taisto's only criticism is the seriously reduced rear visibility in the WA1200; he says there are good reasons for the warning stickers that tell people to stay at least 30 m from the machine when it is working. Operators admit that there will always be blind spots on such big machines and while fitting cameras and extra mirrors certainly help, they still have to take great care.

As previously stated, the Talvivaara mine will not be happy with anything less than 6,000–7,000 hours of operation a year from their WA1200; they also expect it to have a service life of at least ten years. To ensure that it meets the high number of annual hours they have a one-year service contract, with the option of a two-year extension. The Finnish importer stores €200,000 worth of parts at the mine, and any parts not held on site are quickly sourced from Komatsu's Belgian parts facility. 'We have to work hard to ensure that it is always available for operation,' says Seppo.

During the next ten years Komatsu expects to overhaul most components of the WA1200 at least four times. The hydraulic pumps will be changed at 10,000 hours, followed by an engine re-build at around 12,000 hours. The next big job will be the transmission revision at 16,000 to 18,000 hours and then the axles at 20,000 to 23,000 hours.

Big Hitachi Haulers

Not only has Talvivaara set the record with their two big excavators and massive wheeled loader, they also operate eight of the biggest Hitachi rigid dump trucks in Europe. Caterpillar and Komatsu have always tended to have a stronghold on the European truck market, but Hitachi has remained undeterred. Some European mines and quarries already operate with the 65 tonne capacity EH1100, which is the company's most popular truck by volume. A few customers even use the larger EH1700.

Hitachi now has a new range of Japanese-made trucks that they believe will be real contenders for domination of the hauling market. The first of these is at Talvivaara. Called the EH3500ACII, it has a payload of 168 tonnes and is designed to compete with the trucks such as the Cat 789 and Komatsu 730E. This new Hitachi truck is so large

4.20. The Talvivaara nickel mine has eight of these Japanese-made Hitachi EH3500ACII dump trucks, the only ones in Europe.

that seventeen vehicles were needed to bring the parts of each truck to the mine.

TMC considered all the viable options and, in the end, concluded that there was little difference between the purchase prices of the Hitachi in relation to other heavy-duty trucks. They chose the Hitachi EH3500ACII because it offered the kind of reliability, performance/output and local support that was needed on the site. Another big feature in the Hitachi machine's favour was its electric drive-line, which they preferred to the usual mechanical-drive function. 'It is proven technology,' says mine manager Arto Suokas. 'There are no wearing parts and maintenance costs are lower.'

Despite the attractive price and that electric drive-line feature, the EH3500 was a difficult machine to sell because, technically, it did not exist at the time that Talvivaara were looking. The Finnish Hitachi importer had to convince the mine of the truck's compatibility with the EX3600 excavator. Furthermore, as there was no previous track record of any other excavators and trucks at the mine, because it was a greenfield site, it was particularly difficult for Hitachi to convince TMC that a non-existent truck would suit their needs perfectly. Everything that Hitachi did in order to prepare for the new excavators and trucks had to be from scratch and the manufacturer had to

make lots of calculations to ensure it was the best combination. The Talvivaara mine is due to run for another thirty-five to forty years, so Hitachi did not want to risk losing the opportunity for future business.

The first of the larger-than-average haulers was also the very first EH3500ACII delivered anywhere in the world, having been commissioned in August 2008. Most electric truck makers use electric motors from Siemens or Mitsubishi but Hitachi claims to be the only company to have developed their own in-house electric-drive technology for the new truck. When Hitachi completed the first twenty per cent interest in Euclid in 1994, it acquired the R190 – a DC driven truck – which was then renamed the EH3500DC and taken out of production in Canada in 2004. The company then spent the next four years re-engineering it in Japan with the result being the first AC electric drive truck with a nominal payload of 168 tonnes.

Though it may appear a tedious task, the choice to re-engineer can be easily explained by the fact that DC motors require extensive and regular maintenance. Contactor tips and traction motor brushes need replacing at 2,500 hours; the commentator turn and undercut need to be overhauled at 7,500 hours; and the wheel motors need overhauling at 25,000 hours. With an AC system, maintenance is reduced to replacing the main contactors at 7,500 hours and a wheel motor overhaul at 30,000 hours. A further benefit of moving to AC drive is superior braking performance – something that current EH3500DC operators will appreciate. AC drive allows a fully loaded truck weighing over 300 tonnes to be brought to a complete standstill without the use of the service brakes. The DC version is only able to slow it – not stop it. The biggest advantage of AC is that it uses no wearing mechanical parts to stop it but, instead, employs a full electric retarding system.

Another incredible benefit of AC drive is that the hauler can be held on an incline with just the retarder. Operating a bit like an automatic car, this is something that could not be done with the DC truck. It may be obvious, but Hitachi is no stranger to electric drive systems. Since 1992 it has sold 180,000 of them and it boasts some prestigious customers, including the Japanese bullet train and, more recently, a deal to supply new passenger rail rolling stock for the UK. As it turns out, there are lots of common features among electric motors so many of the parts used in the UK trains are also be found in the large trucks.

The electric-drive train update is not the only change to the EH3500ACII because the engine is also brand new. Although it is still a Cummins, the former Tier 1 K2000E has been replaced by the same Tier 3 QSK engine as the EX3600. It generates a maximum output of 2,000 hp in the Hitachi truck. The frame is also new; Hitachi opted to replace the outdated welded sections and this has created a much more streamlined design. This design allows for a new tipping body, which is deeper at the front and provides for an improved weight distribution. The angle of the tail chute has been increased from eight to fourteen degrees to help reduce spillage on the haul roads. A further big feature concerns the front axle suspension system. Although the hydraulic strut design has been retained from the former Euclid truck, they are now easier to remove for maintenance. Perhaps the biggest change, however, concerns the new trailing arm suspension. This suspension not only leads to better stability and ride comfort but also increased tyre life because it minimises the lateral forces.

The inside of TMC's truck bodies contains an additional 13 tonnes of wear plates, which reduces the payload to a

4.21. When fully loaded, the EH3500ACII is capable of carrying a maximum of 168 tonnes of material; however, Talvivaara's trucks carry less because the inside of the tipping body hides 13 tonnes of wear plates.

4.22. Talvivaara is lucky to have a workshop big enough to accommodate these 7 m tall trucks. No mean feat considering that the tipping body is able to rise to 13 m – taller than an average house – and the truck is over 13 m long.

maximum of 155 tonnes; this means that four buckets from the EX3600 are sufficient to fill each one. TMC is hopeful of 60,000 hours and ten years of active service from each truck, during which time there will be two rebuilds.

Although pleased with initial performance, drivers complained the cabs were too cold during the winter. This is despite the fact that the mine ordered the winter package that is commonly specified for Canada. 'Canadian operators must be really tough or else their winters are a lot milder than here,' jokes Arto. The way the system is designed, there should be plenty of heat but the distribution is poor. Also, the operators have to sit just 10 cm from the steel fire plate at the back of the cab where there is not only no insulation but noise and heat were also issues. The Finnish dealer has resolved this problem by adding extra heating pipes and improved insulation of the doors and rear of the cabs on all six trucks.

CHAPTER 5

Spain

The vast majority of foreign tourists who visit Spain each summer come for the beaches and the promise of endless sunshine. A few also head inland to taste the more tranquil side of the country, while an even smaller number come face-to-face with one of the country's biggest natural resources – coal.

As the largest coal producer in the EU, Spain's coal mining industry is big business. During the past decade many millions of euros have been spent on obtaining some of Europe's largest earthmovers to excavate the estimated reserves of 560 million tonnes. However, while coal still generates thirty per cent of the country's electricity needs, the big question concerns how much longer this will last.

Spanish coal is expensive to produce and until recently coal mines and power stations were state-subsidised. This is a contentious issue among European politicians, who continue to discuss the merits of state-subsidised mining. On the one hand, environmentalists are having a field day trying to shut down what they refer to as 'mountain-top mining'. They claim that the subsidies are better spent on new wind and solar energy solutions. On the other, the coal fields in the north-east and west of the country provide badly needed jobs and are crucial to the survival of many rural communities.

In the autumn of 2010 the Spanish government secured the approval of the European Commission to grant its struggling coal mines a lifeline. Cash handouts were to be administered until the end of 2014 – with a glimmer of hope for a stay of execution until 2025. However, at the time of writing, the European Court of Justice had issued a temporary injunction preventing the Spanish government from subsidising the use of domestically produced coal to generate electricity.

As a result, many mining operations have ground to a halt, with companies claiming that without financial help they will be unable to cover operating costs of their big machines. It is clear the Spanish government faces some tough calls because it has no wish to swell the ranks of the four million-plus dole queue.

Ruta de la Antracita

These present-day issues are in stark contrast to an industry that provided jobs for at least 50,000 people several decades ago. There is no better place to witness both past and present-day coal mining activities than in Fabero, a town with a population of 5,000 people which is located in the remote north-west province of León. The nearest major settlement of any significance is Ponferrada, a town that straddles the A6 motorway linking the capital with La Coruña on the Galician coast. Although Ponferrada is the capital of the Bierza region, travellers who do stop here prefer the tranquillity of the delightful village of Molinaseca. Just a couple of miles away from the town but a million miles in all other ways, Molinaseca is best known for three things – a

5.1. The town Fabero has a rich coal-mining history. Even today, the mountains behind the town conceal one of the largest coal mines in Spain.

5.2. The Camino de Santiago de Compostela attracts thousands of tourists each year but most are blissfully unaware that the mountains in the background hide some of Europe's largest earthmovers.

5.3. Coal haulers transporting material from the Gran Corta de Fabero coal mine (seen here in the background) are a common sight on the road from Fabero to Ponferrada.

5.4. The Ruta de la Antracita provides tourists with a flavour of coal mining from the past and acts as a link between former deep mines.

regional meat dish called Botillo; its Mencio grapes, which produce a great-tasting red wine; and as a stopover for pilgrims hiking or cycling the Camino de Santiago de Compostela.

All roads heading north from the bustling town of Ponferrada ultimately lead to the high peaks of the Picos de Europe mountains – Spain's first national park. Heading north from Ponferrada there is plenty of evidence of coal's former importance to the region. The sides of the valleys are littered with old workings of former underground operations. There is also plenty of evidence that coal mining continues today because the roads are frequented by a constant stream of trucks shifting coal to one of two local power stations. The rugged terrain conceals a number of active washing plants, one of which – at 700 m – is at the entrance to Fabero, and the surrounding mountains hide many millions of tonnes of coal.

Coal has always played a vital role here, and the town quickly flourished amidst the rich pickings. Although Fabero's deep mines have long been mothballed, the community is proud of its coal-mining history. Tourist

5.5. The 275 m deep main shaft at Pozo Julia has long been back-filled. At one time over 3,000 people – half the population of the town – used to work here.

offices in the region have joined forces to create a route, known as the Ruta de la Antracita, for visitors keen to learn more. The route showcases Pozo Julia, the underground coal operation opened in the 1950s which used to provide jobs for 3,600 people until its closure fifteen years ago.

Today, coal trolleys and buildings still litter the landscape as do scores of pylons that were once used to carry coal down the neighbouring mountainsides. Fortunately for the town, a few hundred inhabitants still work in Fabero's present-day surface coal mine. The owner of the current operation is Victorino Alonso; he also owns a national construction company (FERPI) as well as eight big surface coal mines, the 1,000 hectare site of Gran Corta de Fabero being the largest.

Operating as UMINSA (Union Minera del Norte SA) and Corto Minero Cantabrico (CMC), Victorino Alonso is no stranger to grabbing the limelight. With a machine fleet numbering in the 3,000s, Alonso's operations utilise some machines not found anywhere else in Europe. The fleet includes a hundred Tamrock drill rigs and approximately 400 dump trucks! Each year, his coal mines supply power stations with 3 to 4 million tonnes, an amount which is roughly equal to thirty or forty per cent of the 10 million tonnes of coal excavated in Spain. It probably goes without saying that this operation relies heavily on some truly massive earthmovers.

Big Cats

Caterpillar's announcement of their return to the heavy-line excavator business has been seen by some as a long overdue decision. Others harbour a more negative outlook, believing that the manufacturer will really need to step up their game this time round.

Over the period 1992–2004, a vast quantity of Caterpillar models, including the 5080, 5090, 5110, 5130 and the 5230, were produced and shipped to various locations around the world. Specifically how many of each model came to work in Europe is an unknown figure; what is known, however, is that the largest Cat excavator ever made – the massive 5230, weighing 320 tonnes – never made the journey across the Atlantic from America.

Despite never having obtained the massive 5230, Europe has been home to a number of 5130s, four of which made it to Britain. Sadly, all but two of these have disappeared; the assumption being that they have been exported. The third machine stands idle in Scotland and

the fourth – a B series model – acts as the prime mover at the Tarmac aggregate quarry in Swinden, North Yorkshire. (See Chapter 6.)

In mainland Europe, a twelve-year-old 5130B recently, and rather randomly, surfaced in the Netherlands where it had been utilised as a prime mover for a German quarry. The machine is likely to return to the US, possibly to support an existing fleet in the Appalachian coal fields. Otherwise, the giant excavator will serve as a sacrifice to keep other machines running a little bit longer. Another 5130B was also recently exported from the Netherlands and modified into a high-rise demolition monster. It currently cuts up disused oil rigs in Norway.

In 2000, Caterpillar received the biggest single European order for the 5130B Caterpillar excavator ever recorded. The order came from the Spanish Victorino Alonso Group and they wanted thirty-two excavators – specifically eight 5130Bs, each weighing 185 tonnes, and twenty-four 5110Bs, weighing 130 tonnes! Even Caterpillar must have been impressed by such a large order.

Before the turn of the century, Victorino Alonso's mining operation relied heavily on contractors. After some serious calculations, however, the boss decided it would be cheaper to invest in new machines and do the work within the company instead of outsourcing. Four years after the immense purchase of Cat 5130Bs, half had to be relieved from their excavating duties by much larger Hitachi machines. Apparently, the largest Caterpillar excavators in Europe did not fit in with ambitious plans to take coal outputs to new, higher levels.

Mine engineer Roberto López Braña explains there was also another issue at the time – the design of the 5130B, which he reckons is too high and too narrow. 'The shape results in quite a bumpy ride so the excavators were not very popular with my operators.'

The 5130ME was Caterpillar's very first 5000 series excavator. In 1992, the company began production on both backhoe and face shovel options. While this machine was no match for the 320 tonne 5230, it was still a serious digger. The 5130ME had an operating weight of 180 tonnes and at the front it carried a 10 cu m bucket. When the 185 tonne B series excavators were introduced in 1997, the initial 755 hp engine was swapped for a more powerful 800 hp variant. Exact numbers are hard to predict, but it is believed that Cat made several hundred of these slightly heavier

5.6. The 5130B was the second largest excavator in the Cat 5000 series. Of the eight bought by Victorino Alonso in 2000, six are still available for front-line work and two have been cannibalised.

5.7. Spanish operators have never liked the rough ride that they have to endure when working the Cat 5130B; however, the fact that they are still going confirms the almost indestructible nature of these excavators.

5.8. A few years ago, the future of the 5130Bs was uncertain but with new mines opening up,
six of the eight excavators are back in action despite having clocked over 25,000 hours each.

machines. The B series machine was produced at the Decatur plant in Illinois and was also highly successful in the Appalachian coal mines, where the mass excavation configuration was able to dig selectively to remove overburden from the top of a thin coal seam without contaminating it.

When the first Hitachi excavators started to arrive at the Spanish coal mines the future of the Caterpillar 5130B seemed uncertain. Some have been relegated to the role of back-up, while the carcasses of others have been disassembled for the parts. Fortunately for the big Cats, Victorino Alonso decided to raise outputs at Gran Corta de Fabero, as well as at the other mines in Tormeleo, Cerredo and Carrasconte. In addition, they have recently opened two new operations at Ponfria and Feixolin and are waiting to make a start on a new mine at Villares. With all these locations to mine, it does not look as though the Cat 5130B can be completely abandoned by the Victorino Alonso Group just yet.

Now UMINSA's eight active surface mines need to deliver 2.5 million tonnes of coal, a fact which makes the brief of mine engineer Roberto quite simple: he will move mountains, just as long as he can deliver the coal! Considering the huge volumes of overburden and thin coal seams, this huge target is a tough challenge, and one which requires as many active machines as possible.

Thankfully, six of the eight 5130Bs have been resurrected to help meet the company target as well as fifteen of their active 5110Bs, all of which have clocked over 25,000 hours. With all the hard work going on, it is perhaps not surprising that it is easy to find evidence of the cannibalisation of exhausted machines, particularly near the entrance to the mine workshop.

Smaller than the 5130Bs, the 5110Bs are ideal for squeezing into tight corners to mop up the remains of a blast. With twenty-four years of service as the Fabero mine's workshop manager, Javier Vasquez knows the Cat excavators inside out and confirms their indestructible nature. 'Both models were certainly built to last and just keep going.'

However, the Cat 5130B is not totally without issues. A couple of slew rings needed replacing after just 8,000 hours, which they put down to the high and narrow design, and the machines had their fair share of cracked track pads. One excavator even needed a new engine after just 4,000 hours.

By comparison, there have been very few problems with the 5110Bs and the mine operators reckon that by utilising parts from some of the older machines they may be able to keep them working for another ten years, long after the anticipated expiry date of the 5130B.

5.9. In the material above Gran Corta de Fabero three 5110Bs with 700 hp engines are the preferred choice for opening up the extension.

5.10. Victorino Alonso bought twenty-four Cat 5110Bs a decade ago and it is quite possible some of them will still be working in 2020.

While many in the industry consider the 5000 series to be one of the worst decisions in Caterpillar's long business history, Victorino Alonso's experience confirms that they were built to last. In fact, used 5110Bs and 5130Bs are now highly sought-after excavators because of their proven durability.

Hitachi EX5500

The Victorino Alonso group bolstered its machine park yet again with a mind-boggling order at the 2004 Bauma show in Munich, Germany. The staggering deal for 150 dump trucks, 180 excavators and 250 wheeled loaders, remains one of Hitachi's biggest single orders ever. The record-breaking order also included five of the largest Hitachi excavators ever brought to Europe – the 520 tonne EX5500.

Critics initially argued that the Hitachi EX5500 excavators were simply too big for this type of mining work. However, these same critics have since been silenced because the Victorino Alonso fleet has been successfully bolstered to twelve machines, which have been spread around the group's eight big coal mines.

The last of the EX5500s to join the Spanish fleet had to stop in Munich first, at the 2007 Bauma show. Easily the biggest excavator on display, it was photographed by tens of thousands of visitors. When the show closed, the task of dismantling the excavator began. Although Hitachi now has plenty of practice, it was still a long 800 km road trip to the mine from the arrival point at Barcelona, and twenty vehicles were required to shift the components of just one machine to one of UMINSA's coal mines. The process then took another two weeks, with the help of two large cranes, to reassemble them. After a further week of tests they were ready for action.

The size of an EX5500 is difficult to comprehend. In fact, the machine is almost as heavy as a new A380 Airbus, a machine capable of carrying over 500 passengers! The hydraulic excavator bucket fixed to the EX5500 is 29 cu m, which is definitely big by European standards. The fact that four of these monstrous machines work at Gran Corta de Fabero is truly incredible.

5.11. In 2004, five Hitachi EX5500s were purchased by the mine. A few years later another seven followed, making this the single largest fleet in the world.

5.12. The EX5500's monstrous arm has no problem reaching over the side of the Cat 789 to empty the contents from its 29 cu m bucket. The excavator can fill the truck in three to four passes when emptying 55 to 60 tonnes at a time.

5.13. The EX5500's 29 cu m rock bucket is huge. It stands 3.5 m tall, over 4 m wide and weighs nearly 30 tonnes even when empty.

5.14. Once the overburden has been removed, a fleet of smaller Hitachi excavators moves to mop up the coal.

Spain's largest coal mine is well accustomed to receiving visitors, especially since aerial television images of the mine were broadcast internationally when the Vuelta – a famous Spanish road cycling race – was featured around Fabero.

On the ground, Gran Corta de Fabero is not short of visitors and enthusiasts either. The largest Hitachi excavators in Europe also have magnetic powers, drawing visitors from all over the world, some of whom have surprising comments. A group of senior managers from an American coal mine once asked where all the coal was! By comparison with American coal mines, the thirteen coal layers at Gran Corta de Fabero are thin and difficult to spot. The largest is a good metre tall, but with some others they have to be satisfied with just 10 cm of coal. Separating these wafer-thin layers are deep wedges of extremely tough rock varying in depth from just 4 to nearly 30 m; the intricacy of the coal-mining operation at this Spanish coal mine displays clearly the need for high-quality machines.

When the mine opened at the end of the 1970s annual coal production was just 300,000 tonnes. Today, the daily target is 4,000 tonnes and they are disappointed if they cannot achieve a million tonnes a year. The EX5500s have their work cut out for them because for every tonne of coal extracted they need to shift 40 tonnes of overburden – the equivalent of 1.4 million cu m per month. The cost of working machines this hard can be staggering: for example, each month Fabero's big machines burn 1.5 million litres of diesel!

In the vast expanse of the mine even a 520 tonne excavator can appear quite small, especially from a distance. Onlookers will wonder just how it is possible that these single-armed machines can literally shape mountains. Close up, however, this first impression quickly changes to one of respect as the scale of the machine becomes clear when the distance between the operator in the cab and the ground is measured at approximately two storeys.

Working relentlessly, the EX5500 excavator averages 55 to 60 tonnes a pass. Sometimes they can load as much as 80 tonnes in one pass, and some operators have even claimed to have topped a hundred tonnes! Experience confirms that in ideal conditions it can shift as much as 1,200 to 1,400 cu m of material an hour, but the tough conditions at some of their mines, particularly at Gran Corta de Fabero, means the Hitachi giants struggle and grunt to load just 800 to 900 cu m during the same period.

5.15. *It is difficult to appreciate the sheer size of an EX5500. Even when fully stretched, mining engineer Roberto López Braña is unable to touch the top of the tracks.*

5.16. *The giant excavator dwarfs a fuel lorry. Every day, the 10,400 litre fuel tank needs topping up with nearly 7,000 litres.*

5.17. The EX5500 loads a Cat 789 with a payload of 177 tonnes. The Cat 789 is the second largest Caterpillar truck in Europe and Victorino Alonso's coal mines own sixty-five of them.

A large machine needs a big engine and the two Cummins power packs beating away at the heart of the EX5500 combine to produce 2,700 hp. Working flat out for nineteen hours a day, the EX5500 is drained at an average of 350 litres an hour, and needs topping up with nearly 7,000 litres every day.

Roberto makes no secret of his initial hesitations when the first EX5500 arrived on site. He was familiar with both Cat and Komatsu, but Hitachi was an unknown and previously unused brand and he concedes he was not entirely convinced it was going to work. However, despite a few initial teething problems with the Cummins engines – not to mention excessive bucket wear and a few boom and stick cracks – he admits he is very pleased with the performance of all dozen machines, and describes them as very reliable and virtually unstoppable. The machines work well and deliver good production figures. They each notch approximately 4,500 hours of operation a year. The oldest machines are now nudging the 25,000 hour mark, and Roberto will be disappointed if they cannot double this – not an unreasonable ask considering the price tag is between €4 and €5 million for each machine.

Some critics argue that they bought too many machines a few years back. However, Roberto is adamant that they bought the machines at the right time and at the right price. During the boom years from 2006 to 2008 they were put to good use while many rival mining companies were forced to wait nearly two years for their new diggers.

Even though the oldest excavator has just turned six they are not expecting to have to replace any of them until they have worked for at least ten years. Given that the Spanish coal-mining sector has a guaranteed future until 2014 it would appear that Victorino Alonso calculated correctly. Roberto's advice is to buy some more. The big question is: will they need any more after 2014?

5.18. UMINSA also owns a hundred Hitachi trucks, many of them EH1100s.
The mine enlarged the original 50 cu m tipping bodies to 70 cu m. These trucks only carry coal.

Hitachi EX3600

The Talvivaara Mining Company in Finland operated the first two 360 tonne Hitachi EX3600 excavators in Scandinavia. However, UMINSA was the first to bring the excavator to Europe and, unsurprisingly given their already enormous fleet of massive machines, they soon went from owning one to owning sixteen.

The EX3600 is another excavator with a comfortably secure position in Europe's top ten largest. A good indication of the size of each excavator is that the mines require fourteen vehicles to bring the huge components from the port of Santander in northern Spain. Easily capable of scooping 40 tonnes at a time, the excavator's 22 cu m bucket is used to load trucks with capacities ranging from 90 to 177 tonnes.

One downfall of this big excavator is that it is not really the best match for the smaller truck and three careful passes are needed to fill it. The care taken in filling these small trucks can lead to lost production and it is possible that when the time comes to dispose of these smaller trucks, the preference could very well be for the larger trucks, which the excavator is able to fill in five or six passes.

On the outside, the EX3600 looks slightly different to its bigger brother working not far away. This is not due simply to its size. It is also a cosmetic issue – there is a grey stripe around the rear of the upper structure of the EX3600. This is because it is a Dash 6 version. Although new EX5500-6 machines now also get the new livery, the twelve machines in Spain are all Dash 5s and so they do not bear the new stripe. Beating away at the heart of the EX3600, the single Tier 2, 16 cylinder, 60 litre Cummins engine produces a maximum output of 1,944 hp. This is a slight improvement on the previous 1,900 hp offered by the EX3500-5.

The EX3600 is covered in Chapter 4. To recap, its vital statistics are impressive. The machine stands over 7.5 m tall to the top of the cab and is fitted with an upper structure that is 9 m wide, meaning there is plenty of head room to walk right underneath the 40 tonne rear counterweight.

Once again, Roberto is very clear on the role of his big fleet of EX3600s, which are expected to be available for two 10-hour shifts per day and at least 4,000 hours a year. In order to perform in this way they use an average of 250 litres of diesel an hour. Fortunately, the 7,450

5.19. The eight surface mines owned by Victorino Alonso are the domain for sixteen Hitachi EX3600 excavators – easily the biggest fleet in Europe. Each of these hydraulic monsters weighs nearly 360 tonnes.

5.20. The 10 m backhoe boom, 6 m stick and 22 cu m bucket account for almost 83 tonnes of the total operating weight of the excavator. Each time it empties its bucket, roughly 40 to 45 tonnes cascades into the truck bed.

5.21. Behind this EX3600 is one of the massive EX5500s. The combined efforts of some of the largest Hitachi excavators in Europe are slowly shaping the landscape at Gran Corta de Fabero.

litre fuel tank – which is 250 litres larger than the EX3500-5 – provides ample capacity for a full day's operation.

Although the EX3600 offers slightly faster working speeds than the EX5500s, things got off to a bad start following a plague of electronic problems. These centred on the more sophisticated nature of the EX3600s. Containing higher levels of electronic wizardry then the EX5500s, these delicate systems tend to be sensitive to dust and vibration, and Roberto claims a fire broke out in the electronics department of one of their excavators. He is also concerned that the electronics could be affected by their climate, which can see temperatures change quickly from minus thirty to plus thirty degrees Celsius. 'Electronics are great in the office but not in a mine,' he adds. 'We like simple mechanical machines.'

Komatsu D475A

When it comes to removing overburden, brute force is what is initially needed but the big excavators have to stop about a metre from the coal seam. The final layer is then carefully pushed off using a dozer or extracted using a much smaller excavator.

The Victorino Group runs a fleet of twenty Komatsu D375A dozers, but all of these are dwarfed by the Dash 3 version of the D475A, which is the largest Komatsu dozer in Europe. This machine is second only to the D575A –

the world's largest production dozer. First manufactured in 1986, the D475A-1 was a 740 hp replacement for the 650 hp D455A. Although originally there were a couple of D475As in France and Germany, these days Spain is quite possibly the only country in Europe where they can be found.

Although it might not look large, the dozer at Gran Corta de Fabero is nearly 12 m long and stands almost 5 m tall. With an operating weight of over 110 tonnes, power comes from a 30 litre, 12-cylinder Komatsu engine that peaks just short of 900 hp. The front blade is 5 m wide and has no problem pushing more than 30 cu m of material, while the rear ripper – weighing in at 7 tonnes – is useful for digging down to depths of nearly 2 m.

However, after over 30,000 hours of punishing duties, the old Komatsu dozer is currently out of action awaiting a major overhaul. A new engine, blade, transmission and pumps are just some of the parts that it will be necessary to replace; if the Victorino Group decide to commission the work they could be left facing a €200,000 repair bill. The likelihood that this decision will be made while the future of the Spanish coal industry is so insecure is slim. The future of this giant dozer hangs in the balance.

5.22. Spain is thought to be the home of around ten D475As, which are the largest Komatsu tracked dozers in Europe.
These 110 tonne monsters are easily capable of pushing 30 tonnes at a time.

5.23. Hitachi orange is not the only colour in Victorino Alonso's excavator fleet – the group has plenty of Komatsu experience.
At one time they owned twenty-four PC1100 excavators which prompted them to buy three 260 tonne PC3000s like this one.

5.24. With so many large machines in action, it is easy to overlook the fleet of twenty Cat D10 dozers, which are generally used to scrape the final metre of overburden off the top of a coal seam.

5.25. The Spanish coal-mining giant has always had a good experience with their Demag excavators. At one time, they used to own fourteen H 95s which have since been scrapped. This H 255S will be joining them.

5.26. *After 40,000 hours of punishing high-altitude work, this 1996 Demag H 185 recently relinquished its prime-moving duties. It now stands idle, its fate undecided.*

Komatsu PC2000

Gold and silver were once the most highly sought after minerals in Europe, perhaps in the world. Today, however, coal has become a mineral of equal, if not greater, worth and the Santa Lucia mine in Spain generates an incredible amount of this highly prized substance. Hidden deep inside the mountains, the presence of the mine is only betrayed by the final section of a conveyor belt, which plummets down the mountainside feeding a power station with nearly 3,000 tonnes of coal a day.

Higher up the mountain, a public road passes by the site affording stunning views into the depths of the mine some 500 m below. The operation is owned by Hullera Vasco-Leonesa, which contracts all overburden and coal extraction work to the Spanish company Transportes Peal, which, along with EPSA and FERPI, is one of Spain's top three contractors.

Transportes Peal is a big international contractor that has constructed many things for the region, including motorways, railways, ports and airports. The company began coal-mining activities at Santa Lucia in 1986 from a humble base at the bottom of the mountain. The former premises still exist but the company now operates from much grander headquarters and workshops at Navatejera, south of León.

Until recently, the contractor's extensive machine armoury at Santa Lucia was headed by a Terex O&K RH200. There were also two big Cat 994 wheeled loaders and several Cat 789 haulers, but all have now been shifted to a Mexican copper mine. The Spanish mine's operation and output targets demand the efforts of a select bunch of seriously large earthmovers. These machines are needed to move 8 to 9 million cu m of overburden a year – an average of 45,000 cu m a day – to expose their thirteen coal seams.

Transportes Peal has a fleet of thirty-four 90 tonne trucks which are filled by one of eight prime movers. The largest

5.27. Blasting takes place regularly at the Santa Lucia coal mine because each year they need to shift 8 to 9 million cu m of rock to reach annual coal targets of 600,000 tonnes.

excavator on the site is a Terex O&K RH120E weighing 282 tonnes. The rest of their loading fleet is all Komatsu – a PC1100-6 weighing 115 tonnes and two PC1250-7s.

However, the star attractions of the current Komatsu fleet are a pair of PC2000-8s that were at the time of their arrival the very first ones in Europe. Bought nearly three years ago, Transportes Peal has obviously been pleased with their durability and work ethic because they recently purchased two more for one of their nickel mines near Seville.

The design for the PC2000-8 has developed and matured over many years. Although it replaced the PC1800-6 at the 2007 Bauma, this model was actually spawned from another PC fledgling, the PC1600 from 1998. In fact, some of the ideas for the original PC2000 model came from an excavator known as the PC1500-1 which was around in 1982.

5.28. This Terex O&K RH120E is the largest machine on site. After nine years and 40,000 hours of operation its 17 cu m bucket is still capable of loading upwards of 550 cu m per hour.

5.29. Blending well with the background, Transportes Peal's Komatsu excavator fleet at Santa Lucia includes two PC1250-7s with 7.5 cu m buckets.

5.30. With coal reserves at the bottom of the mine nearly exhausted, this Komatsu PC1100 is opening a new push-back at the top of the mine. It has an operating weight of 110 tonnes and carries a 7 cu m bucket.

5.31. Two of the Komatsu PC2000s owned by Transportes Peal work at Santa Lucia, each with an operating weight of 200 tonnes. The 13 cu m bucket fitted to the front of this machine is not the largest the excavator can carry.

With an operating weight of 200 tonnes, it is difficult to know whether to describe the newcomer as the smaller sister of the 260 tonne PC3000 or as the bigger brother of the 110 tonne PC1250. Whichever way this machine is classed, however, its weight and the fact that it was created alongside the PC3000 and PC4000 in the Komatsu plant in Osaka, Japan, clearly puts it in the mining sector.

The Japanese have high hopes for the PC2000-8 and are confident that this weight of excavator will enjoy a bright future. An outlook as optimistic as this should be fairly well founded. It would appear that the Japanese have reason to congratulate themselves because they have created an excavator that is perfectly matched for filling their most globally popular dump truck – the 91 tonne capacity Komatsu HD785.

The main role of the two excavators working at Santa Lucia is to strip overburden from new push-backs at the top of the mine. The flexibility of the PC2000-8 means that it can also be quickly diverted to benches lower down in the void when extra power is needed.

According to production manager Juan Ramón Garcia, during the week all machines on site are expected to work round-the-clock and the PC2000-8s both work approximately 4,000 to 5,000 hours per year. While the machines at Santa Lucia do not work weekends, Saturday is a particularly important day for giving the machines a thorough clean. Although they only stay clean for a day at the most, this provides the service team with an ideal opportunity to give each machine a thorough maintenance check.

5.32. Komatsu is certain that the PC2000 will have a bright future because it is a perfect four-pass match for filling the 90 tonne capacity HD785 – Komatsu's most popular mining truck.

This is exactly what Komatsu had to do when the PC1800-6 became the PC2000-8; there have been so many upgrades that the machine is virtually unrecognisable. Engineers were given a challenging brief: not only to boost performance and improve operator visibility and comfort, but to package it all in a much simpler design. The result is a powerful but neat and compact design that ticks all the right boxes for the Spanish customer.

On the outside, give or take a few centimetres, the vital statistics are not too different from the former PC1800-6. However, the top rear half of the upper structure is where the similarities end. Not only does this part of the machine look completely different but it is nearly 1.5 m wider. The difference in physical appearance is partly due to the new design of the rear counterweight, which is just half the height of the previous version. As a result, engineers were able to

remove the access stairway to the top of the former counterweight and replace it with a catwalk, anti-slip plates and handrails around the complete rear of upper structure. These changes have made the whole area safer for the operator.

The larger cab is also new. The idea was originally used on Komatsu's larger PC mining excavators and when applied to the PC2000-8, it succeeded in reducing operator noise levels from 71.3 decibels, the recorded figure in the old PC1800-6, to 66.8 decibels. Overall, Komatsu has made a huge effort to improve things.

The biggest single change was also the starting point for developing this excavator. Europe, Japan and the US were in need of new engine technology that would comply with emissions targets. Komatsu's solution was to ditch their former twin-engine design: two 6-cylinder 15 litre engines which when combined were capable of producing 900 hp at a rated 1,800 rpm. As a

5.33. Launched in 2007, the PC2000-8 not only looks completely different from the PC1800-6, but it also features a complete interior upgrade. The former twin-engine design has been ditched in favour of a single 12-cylinder 976 hp power source.

replacement they introduced a single 30 litre, 12-cylinder Tier 2 version rated to 976 hp at the same engine speed. This engine has been tried and tested by the Komatsu HD785 truck and D475A dozer and has been deemed worthy of the excavator.

Not only have the number of engines in the PC2000-8 been reduced, so have the hydraulic pumps – only two remain of six – and the travel motors, which have been halved from four to two. There is now just a single radiator and oil cooler where there had previously been two and further new technology has been introduced to reduce noise levels. This technology comes in the form of electronically controlled variable speed fans, dubbed 3D-Hybrid, which reduce inflow air turbulence.

Swing and travel speeds remain the same at 4.8 rpm and 2.7 km/hr respectively, and while the stick digging force remains unchanged at 586kN, bucket forces are up from 688kN to 697kN. This is not simply due to 76 hp more engine power and a slight increase in maximum hydraulic oil flow rates from 2,304 to 2,317 litres per minute, but also from a complete redesign of many vital front-end components, including new positioning and larger hydraulic cylinders.

Just like the PC1800-6, the standard bucket for the PC2000 remains at 12 cu m. However, the newer model can also carry a 14 cu m bucket. The excavators owned by Transportes Peal each wield 13 cu m buckets which can fill the 90 tonne capacity trucks in four passes. In the right conditions each massive excavator can load anywhere from 450 to 500 cu m an hour.

Considering the size and production figures at Santa Lucia, experts question why Transportes Peal opted for the PC2000 and not the larger PC3000, the favourite choice for most European mines. Juan Ramón thinks that the purchase of the PC2000 model gives the company the option to use them in the public works sector. 'A 260 tonne excavator is not only too large for many motorway projects, but on a basic level, it is also too big and bulky to transport easily.'

While Santa Lucia deposits contain enough coal to last at least another twenty years, Transportes Peal has obviously kept its options open. With the plight of the Spanish coal sector a definite concern, the survival of coal mines after 2014 is far from secure; therefore, it would seem sensible for the Spanish contractor to have plenty of alternative uses for its PC2000s if it does all go pear-shaped.

5.34. Ariño has it all from dinosaur footprints and cave paintings to the Santa Maria lignite mine just up the hill. With an annual production of over a million tonnes of lignite a year, Santa Maria is one of the biggest mines in Spain.

Ariño

In the Spanish province of Teruel, from the 1940s to the 1970s, soft coal (lignite) used to be excavated from deep mines found in the rugged, mountainous interior. Situated between Madrid and Barcelona, the region is now celebrated more for its cured ham than anything else.

Most deep pits closed many years ago and the history of the region's mining endeavours can be seen through a visit to the MWINAS mineral park on the outskirts of Andorra. The park provides visitors with an insight into some of the men and machines from those earlier days: a star attraction is the 43 m head frame and winding gear, that locals affectionately refer to as their own Eiffel Tower!

In town there is also plenty of evidence of former lignite mining activities, sights which are in stark contrast to the chimneys of a modern-day, coal-fired power station protruding above the skyline in the distance. The modern power station is actually one of the largest in the region and most of its lignite is supplied from one major mine – Santa Maria. From Monday to Friday a constant flow of trucks can be seen leaving the washing plant at the mine and heading for the power plant.

With headquarters in Teruel, in the north-eastern province of Aragon, Santa Maria is not only the largest of SAMCA's four surface mines but also currently one of the biggest operations in Spain. Located in the centre of a triangle formed by Zaragoza, Barcelona and Teruel, it is within sight of the tranquil mountain village of Ariño.

In early October the parched scrubland in the surrounding area is in desperate need of some autumn rainfall, despite the fact that just a short drive north, the mighty Ebro river provides an abundance of irrigation water for neighbouring farmland. However, further south the lush vegetation quickly gives way to a plateau, ultimately rising to become the Sierra de Arcos, a rugged outcrop popular with hikers.

This is a region of almond trees and olive groves, while higher up sheep flocks are herded through the barren landscape – a process that has remained largely undisturbed for centuries. Evidence of the region's historic past can be confirmed by numerous cave paintings, many of which have been dated to 6,000 BC.

5.35. Every week day 280 fully loaded trucks carry 10,000 tonnes of lignite to the power station.

However, the cave paintings are not the reason for the village; like so many other mountain villages in this remote area, Ariño owes its existence to the lignite buried deep below the surface.

The local community is clearly proud of its mining past, a fact which can be seen from the relic of a head frame and winding gear from an underground mine located next to the main through-road. Travellers with time to stop and walk the narrow and very steep streets will find a couple of old coal trolleys among other artefacts. There is even a street called Calle Minas,

which does eventually meander up the mountain to the last remaining deep pit – providing spectacular views of the surface mine on the way.

Most lignite is brown in colour but Santa Maria's lignite is unique, taking on a deep, dark black similar to the night sky found in this remote region. The deposit was formed 100 to 120 million years ago in a river delta and experts have estimated that there are reserves of lignite that will last well beyond 2025. Sandwiched between layers ranging from 50 cm to 5 m, the deepest reserves lie at around 500 m. Unfortunately, the

5.36. All lignite extracted at Santa Maria ends up at the Andorra power station, one of the largest in the country.

5.37. The mountain village of Ariño is best known in Spanish mining community as the home of one of the biggest coal-mining operations in the country.

economic limit of the surface mine is 250 m, so further exploitation of the deeper layers has to be tackled from underground tunnels.

The Sierra de Arcos is a palaeontologist's paradise and just outside the village of Santa Maria are a series of well-marked dinosaur footprints close to the Río Martín. The area's ancient past is best highlighted by the extensive numbers of fossils and skeletons, most of which are on display at the Dinópolis Theme Park at Teruel. The park's panoramic museum reportedly contains one of the world's best collections of dinosaur skeletons – 4,000 of which were uncovered at Santa Maria lignite mine.

When the mining activities have finished, there will be the obvious question of what will need to be done to restore the land. In the past, the Spanish mining industry suffered from a poor reputation when it came to green issues. Today, many mining companies are doing all they can to limit their impact on the environment. Working in close cooperation with university students, the goal for the Santa Maria site is to restore the land to as near a natural state as possible, and then plant it with native plants and trees to create a wildlife haven.

Liebherr R 995

In order to reach the lignite seams at the Santa Maria mine, a staggering 80 m deep layer of some of the most colourful overburden imaginable has to be removed. Over 13 million cu m of limestone, clay and sandstone is shifted by big machines in just one year.

Two decades ago, this was the task of five Demag excavators; two H 185S excavators, which weighed 211 tonnes, and three H 185 excavators, which weighed 209 tonnes. Working two shifts per day from Monday to Friday has not stopped these machines from running even after twenty years of operation.

More recently, these five prime movers were joined by five new and seriously large Liebherr excavators. A popular brand with many quarry and mining customers, as a rule their size is limited to the R 984, which can weigh between 120 and 125 tonnes. There is estimated to be at least seventy Liebherr R 984s in Spain.

At the Santa Maria site, there is a contractor's R 984 that is used to remove the last few scraps of waste rock from the top of the lignite seams. When operators need more serious overburden extraction they rely on two

5.38. Twenty years ago, five Demag H 185 excavators were the prime movers at Santa Maria and all five have notched up close to 30,000 hours. They can still be relied on to work.

5.39. At least seventy Liebherr R 984 excavators are estimated to be working in Spain. Although it is a sizeable excavator, at Santa Maria it is limited to cleaning up the scraps left by the larger machines.

5.40. This is one of two Liebherr R 994s that works at the SAMCA-owned mine. Slightly smaller than the R 994B, this model weighs 230 tonnes.

much larger R 994s – each one of which weighs around 230 tonnes.

There is also a single Liebherr R 994B that was commissioned towards the end of 2001. When new, the formidable machine, powered by a 1,500 hp engine, was capable of extracting nearly 1,800 tonnes per hour even in un-blasted material! There are just three of these 296 tonne machines in the country and the other two are owned by Spanish contractor EPSA.

Of all the Liebherrs that work at Santa Maria, the most impressive Liebherr excavator in the country is the R 995: not just the biggest Liebherr excavator in Spain, but also in Europe. Although a handful of these 450 tonne excavators have been married to a pontoon for dredging purposes, the excavator at Santa Maria is believed to be the only land-based example in Europe. In the vast expanse of the mine it is difficult to appreciate fully the immense size of this machine. The R 995 is heavier than a Boeing 747-400 – one of the largest freight-carrying aircraft on the planet – and when fully loaded with 112 tonnes, this aircraft tips the scales at 412 tonnes and travels at speeds exceeding 900 km/hr.

Bought new in 1998, the R 995 was not built for speed, but for one purpose only – to dig. In its prime, this excavator was able to shift more than 2,800 tonnes per hour. It is almost 8 m wide, and the operator sits at an eye level almost 9 m above the ground. The 16-cylinder, 2,140 hp engine is big – even by excavator standards – and easily capable of draining 5,000 litres of fuel during a sixteen-hour working day!

However, the earthmoving duties of Liebherr's largest excavator in Europe could soon be over because after twelve years and more than 30,000 hours of punishing operation shifting millions of tonnes of material, it needs a bit of restoration work – including a new engine. Mine management are currently deliberating whether to take the plunge and invest in a complete overhaul including a new 65 litre MTU engine, or whether to axe it for a brand new R 9350.

With an operating weight of 320 tonnes and an 18 cu m bucket, the R 9350 is a good deal smaller than their R 995. Fortunately, Santa Maria has experienced the R 9350 already, since they own one of the other two R 9350s in Spain and they know it can do the job. The machine was delivered in September 2008; so far it has done over 8,000 hours and has given ninety-eight per cent availability.

5.41. Santa Maria has one of the only three Liebherr R 994B excavators in Spain.
This machine is ten years old and coupled to a 17 cu m bucket.

5.42. With an operating weight of 440 tonnes, SAMCA's R 995 is the biggest land-based Liebherr excavator in Spain and in Europe.

5.43. *The colossal size of the largest Liebherr excavator in Europe is best highlighted by the yellow speck – one of the mine's staff – in the centre of this picture.*

5.44. *A colossal excavator needs a bucket to match and SAMCA's 24 cu m version stands 3 m high and more than 4 m wide. Even empty it weighs nearly 29 tonnes.*

5.45. When the main overburden team has done its work, the lignite is extracted by a Liebherr R 974.

Liebherr fans will no doubt hope that the worn out R 995 will receive a new lease of life and keep going for a few more years. However, the need to produce over a million tonnes of lignite a year and the vast volumes of overburden that need to be moved to expose it, may force the mine operators to favour the reliability of a brand-new R 9350.

The R 995 is not the only machine that is in need of some attention. The two R 994s at Santa Maria will also need updating at some point because they have notched up nearly 25,000 hours. The Demag excavators still have some life left in them yet, but they will not keep going forever, and mine management may soon need to sacrifice at least one of them to keep the others operational.

Fortunately for the Santa Maria mining operation, there is no rush because it is not necessary for all ten major excavators to work at the same time. The first important decision will concern the future of the R 995.

Big Liebherr Trucks

Santa Maria is the ideal place to see a wide variety of large and very rare dump trucks. Not all of these belong to SAMCA because the mine subcontracts some of the overburden transportation to other companies. On the site the subcontractors can be found using ten Italian-made Perlini DP 705s and several Hitachi EH1100s.

However, all the seriously large trucks – and there are plenty of them – belong to SAMCA. Their fleet includes the HD1200M which is believed to be one of the largest and rarest Komatsu hauler in Europe. Although no longer in production, the Spanish mine originally bought nine of these machines, but after notching some serious working hours, only five trucks remain operational.

The Cummins-powered truck can generate a maximum output of 1,200 hp – fortunate because when filled with 120 tonnes of material the loaded weight tops 200 tonnes. The mine also needs the help of the biggest Liebherr mining truck in Europe – the T 252. This massive diesel-electric truck can be fully loaded with

5.46. With a payload of 120 tonnes, the HD1200M is believed to be one of the largest Komatsu dump trucks in Europe.

200 tonnes of material and when it is it tips the scales at over 330 tonnes.

Sadly only two of the four are operational today. The other two, which were taken out of production after 23,000 hours, are currently being cannibalised in order to extend the working lives of the others.

In addition to the T 252s, SAMCA also has eleven Cat 785 trucks: nine 785Cs and two 785Bs. The burning question now concerns which trucks the mine will buy in the future. Their last five Komatsu HD1200s will not keep going forever. Neither will the two remaining Liebherr T 252s and new replacements for the Liebherr

model are out of the question because the truck is no longer made. While Liebherr previewed the prototype of a TI 274 at the 2010 Bauma show, this 320 tonne hauler remains in the testing phase. For now, the T 282C is the only commercially available truck from Liebherr. With a payload of 363 tonnes, and a fully loaded weight of 600 tonnes, this massive hauler is simply too large for the Spanish mine. Caterpillar would appear to be the likely choice when the time comes to replace Santa Maria's last two Liebherr trucks; for now the mine is keeping its options open. It will be interesting to see which brand name the next new trucks will wear.

5.47. With a payload of 200 tonnes, the T 252 is the largest Liebherr rigid dump truck in Europe. Santa Maria bought four of them – two are now retired and the remaining two are the last T 252s working in Europe.

5.48. Although it is not the biggest Cat truck in Spain, the Cat 785 is still a serious hauler. Santa Maria owns eleven.

CHAPTER 6

England

During the second half of the 1900s, rope shovels and crawler draglines were a common sight in England. One of the first seriously large machines to be found slowly shuffling across the mines of Britain towards the end of World War II was a Bucyrus-Erie 1150-B that worked at Ewart Hill, a site operated by the Parkinson Strip Mining Company.

The next two machines to be imported into the country were purchased second-hand by the National Coal Board in 1953 for the Wilson Lovat site at Terpentwys near Pontypool. The trend was followed with the purchase of two Marion 7800s in 1954, one of which went to Millers' Radar North site in Northumberland.

In 1969, the march of the giant machines took on a new meaning when the 3,000 tonne Bucyrus-Erie, nicknamed Big Geordie, arrived to work at Crouch Mining's Butterwell mine near Morpeth, Northumberland. Big Geordie, the biggest machine of its kind in Western Europe, and the landscape of Northumberland were a better match than could have been predicted.

Big Geordie's bucket was big enough to park three cars in and, as a result, it had no problem gouging out one hundred tonnes of material in a single bite.

6.1. Ace of Spades was a truly breath-taking walking dragline. This picture was taken in 1991 when the 4,200 tonne P&H 757 went to work for the very first time.

6.2. The Ace of Spades had a huge 50 cu m bucket easily capable of shifting 100 tonnes at a time. During the course of its life, it shifted an estimated 300 million tonnes of material.

Despite several unsuccessful attempts to sell the big machine, it was eventually parked at the mine in 1993. Ten years later Big Geordie was sold for scrap.

The replacement for Big Geordie was another walking dragline that was so massive that it had no equal in Europe at the time of purchase. The Ace of Spades, as it was affectionately named by an eight-year-old schoolboy, was a truly massive P&H 757 that weighed approximately 4,200 tonnes. It cost British Coal a cool £18 million to purchase and took eighteen months to assemble at the Stobswood surface coal-mining operation, where it started work in February 1992. For the next decade the dragline's immense 50 cu m bucket – which weighed 80 tonnes even when empty – regularly moved one hundred tonnes of material at a time.

It is not enough to say that the Ace of Spades stood nearly 15 m taller than Nelson's Column. In order to fully appreciate the colossal size of this machine, imagine that it was backed up against the goal posts on one side of Wembley Stadium, then consider that from this position the Ace of Spades was able to throw its excavator bucket into the penalty box at the other end of the pitch: a total distance of 105 m! During the course of its life, the Ace of Spades shifted an estimated 300 million tonnes of material making it not only massive but also powerful.

When British Coal was privatised, the massive walking dragline was acquired by UK Coal. In June 2003, the task of stripping overburden from the lower six coal seams was completed and with no other work suitable for the humungous Ace of Spades, the last truly large walking dragline in Britain lowered its 100 m boom for the last time. Soon after, it was dismantled and transported back across the Atlantic to work for the North American Coal Board at one of their suitably large sites.

Sundew

Sundew, a Ransomes & Rapier W1400 that was originally named after the winner of the 1957 Grand National, is an interesting machine. Although much smaller than either Big Geordie or the Ace of Spades, the reason it is so interesting is because it was one of the few machines that did not work in coal. With an operating weight of 1,675 tonnes, a reach just over 84 m and a bucket capacity of 27 tonnes, Sundew started life in an iron stone quarry in Rutland that was owned by Stewarts & Lloyds.

In 1974, the decision was made to transfer the machine to a new quarry at Corby owned by British Steel. The plan was to dismantle the machine and move it piece by piece to the new location; however, the costs of doing so were estimated at around £250,000, so operators walked the machine the 13 miles instead. The dragline's journey was captured on film by Blue Peter, allowing thousands of young viewers to see how the dragline crossed three water mains, ten roads, a railway line, two gas mains and a river. Nine weeks later Sundew arrived at its new home.

Unfortunately, when steel-making was abandoned at Corby, there was no further need for the massive earthmover. On July 4, 1980, Sundew walked to its final resting place where it remained for seven years before being scrapped over a period of six months, from January to June 1987. Today the only parts of this gentle giant that remain are the cab, which is preserved at the Rutland Railway Museum, and one of the buckets,

which is on display at the Steel Heritage Centre at East Carlton Country Park.

Chevington Collier

Ultimately the super-sized walking draglines lost out to modern hydraulically operated excavators. Smaller and more mobile, the hydraulic machines could be transported around and between sites more easily, as opposed to the large draglines, which were costly and time-consuming to move.

The demise of the walking draglines also had much to do with the size of England's opencast sites. Once the large coal seams were exhausted many remaining operations were no longer big enough to cope with the huge appetites of the massive machines. Even smaller draglines like the Chevington Collier – a 1,288 tonne Bucyrus-Erie 1260-W that worked at Steadsburn surface mine on the North Yorkshire border – needed a cut of at

6.3. Until recently the Chevington Collier, a 1,288 tonne Bucyrus-Erie 1260-W, was one of the largest remaining walking draglines in Britain. It used to work at the Steadsburn coal mine owned by UK Coal.

6.4. In its prime, the giant 32 cu m bucket of this dragline – which was suspended at the end of 54 m long boom – was capable of scooping up 43 tonnes at a time.

least 1 km: something that is no longer possible in Britain.

Made in 1975, the Chevington Collier spent its first few years stripping overburden at a sister site in Northumberland. The giant 32 cu m bucket was suspended from a 54 m high boom and was easily capable of excavating 43 tonnes at a time. Originally purchased by RJB Mining, all accounts suggest that in its prime this machine used to consume electricity at the rate of several hundred pounds sterling an hour! In 1993, the Chevington Collier was refurbished for £1.5 million by British Coal. Then, after only ten more years of operation, it was removed from active service and sold for £250,000. The walking dragline was supposed to join an identical machine in South Africa, but it was never collected. In 2009, the decision was made to strip it for parts and EP Industries dismantled the rest.

Oddball

The days when large Bucyrus walking draglines ripped fossil fuel from the ground of the English countryside have long gone and the speed at which these giant earthmovers disappeared from the British mining scene is nothing short of amazing. Even at the start of the 1990s there were still quite a number at work in the UK, but within the next decade their population had radically declined. One exception, however, is a 1,200

tonne Bucyrus-Erie 1150-B, which is believed to be the largest and oldest surviving machine of that model to be preserved in the UK, perhaps even in Europe.

One of five identical machines that came to the Britain, this sole survivor was built in 1948 at a factory in the south of Milwaukee, in the American state of Wisconsin. For the first four years, it worked at a coal mine stripping operation in West Virginia. During its life it has been moved three times: the first was in the mid-1950s when it was dismantled and shipped to a Wilson Lovat site at Tirpentwys near Pontypool for a job at the National Coal Board's coalfields in South Wales.

In 1972, a fleet of trucks were required to transport the huge machine to its second home at Cannock, in Staffordshire, where it resumed work shifting more than 20,000 tonnes of material a day. Once work was completed at that location, the massive machine was moved on to its third home: St Aidan's opencast coal mine. This operation, owned by Sir James Miller & Partners, was at Allerton Bywater on the outskirts of Leeds.

The newcomer was nicknamed Oddball because of the massive transformers that were needed to convert the 6,600 volt, 50 hertz British electrical power to the American 60 hertz that it required. Oddball was a truly massive machine: roughly the size of sixty double-decker buses. The fact that it was moved not only once, but three times, is incredible.

6.5. Photographed in 1999, this is Oddball, a Bucyrus-Erie 1150-B which weighed 1,200 tonnes. Oddball is believed to be the largest and oldest surviving walking dragline of its type in Western Europe.

At St Aidan's, Oddball was later joined by another dragline – a newer Ransomes & Rapier W2000, which was one of two to arrive in the UK during the early 1980s. This first W2000 was nicknamed the Chevington Lady. The second, and possibly one of the last ever built by Ransomes & Rapier, was erected on site at St Aidan's in 1982.

Their task was to expose what was known as the Barnsley seam. This was achieved by using the veteran Bucyrus-Erie 1150-B to re-handle the material excavated by the newer machine in order to make space for it to expose the lower seams. Working ahead of the two draglines, were a fleet of Ruston Bucyrus rope shovels.

First opened in the 1940s, the opencast mine was extended in 1981 and expected to yield 6 million tonnes of coal over the next ten years. A large chunk of the coal excavated at St Aidan's was supplied to the Ferrybridge power station. At the time it was one of the largest coal mining operations in Britain, but the mine became internationally renowned on Saturday, March 19, 1988 when mining was brought to an abrupt halt by the catastrophic failure of a river bank near Lemonroyd lock. This failed river bank allowed water from the River

Aire to pour into the void and the breach was so large that water even started to flow back upstream. At its peak a staggering amount of water cascaded into the bottom of the void forming a 100 hectare lake filled with nearly 20 million cu m of water run-off.

Fortunately, no one was injured and none of the machines were lost, but it was a mammoth task to reach the remaining 2 million tonnes of coal. To offset some of the costs the site was enlarged to access another 250,000 tonnes of coal. The process of pumping out the water and re-diverting both the river and the canal lasted until 1995. All the coal was eventually removed, but the flooding heralded the end of the road for Oddball and a few other machines. The W2000 was brought back into action to assist in obtaining the last reserves of coal, then, in 2003, when that job was complete, it was scrapped. The fate of Oddball looked to be similar: the machine stood idle for the best part of ten years, moving only once to allow miners to get at the coal lying below her. Fortunately, the colossal prime mover was spared death by gas torch by the activities of the Friends of St Aidan's B-E 1150 Walking Dragline.

The voluntary group is made up of enthusiastic fans

led by Chairman Paul Thompson. He says that they were keen to preserve it as a memorial to the thousands of sunshine miners who worked with iconic machines, such as Oddball, in the coalfields of England, Scotland and Wales. Toward the end of the 1990s, the group made it their mission to scour the country trying to save draglines in disrepair. The group of volunteers even tried to salvage the W-2000 and one of the 195-RBs that used to work at St Aidan's opencast coal mine; in the end, however, they had to make do with the single machine.

With a large generator to supply temporary power, the gentle giant trundled to its final resting place following the closure of the opencast site and the subsequent restoration work. Walking a metre at a time at a top speed of just 0.2 miles per hour, the 49 m move took place in January 1999. Oddball lowered its boom for the last time at the entrance of what is now a wetlands site protected by the RSPB.

With help from a company called Beeby Plant Repairs, trust members carried out a few repairs and painted the machine to restore the original livery and name of the National Coal Board. The machine was officially opened for public access by Richard Budge, chief executive of what was then RJB Mining PLC, on July 8, 2000.

In December 2010, the group celebrated the machine's 62nd birthday with a short meeting in the snow, before retreating to the warmth of a local pub for a celebratory bite to eat and a beer. Other than special occasions such as this, the machine is open to the public for just four days a year. The group has to make the machine accessible for one day in April, one in June and then for two days during the National Heritage Weekend in September because restoration was partly funded by National Lottery money. Receiving an average of a hundred visitors a day, some from as far afield as Ireland and Germany, during the past ten years this popular tourist activity has attracted between 4,000 and 5,000 visitors.

While the carcass of the Bucyrus-Erie 1150-B was salvaged, the gentle giant will never work again. The cost of supplying a suitable power source and replacing the many copper components that were stolen from the electric motors when the machine was parked up is just too high to justify. Sadly, the internal workings were also decimated and making them operational again would cost a fortune. On top of all that, insurance costs and safety issues for a machine of this size were just staggering. 'It is just not practical to run the machine so we will have to make do with what we have,' says Paul.

Even so, a trip to see the remains of Oddball is definitely worthwhile. After all, how often do you get the opportunity to sit in the cab of a machine of this size and get some idea of what it would have been like to wield a bucket measuring 25 cu m? Also included is an opportunity to take a peek inside the belly of this beast, which used to house all the electrical components. Not to mention that the group have even preserved the old sub-station that housed the transformers, which today functions as an information centre. The good news is that it is all free.

O&K RH120

While walking draglines disappeared from the British mining scene many years ago, coal still plays a vital role in generating almost a third of the country's electricity requirements – a figure that rises to almost fifty per cent during the winter. Each year Britain still has to import around 43 million tonnes of coal, nearly 36 million tonnes of which is used for power generation.

More than half the 18 million tonnes of coal produced in Britain each year is recovered from seams lying close to the surface; in many cases, mining companies are purchasing abandoned shallow underground pits, tidying up the scraps and uncovering seams that could not have been reached by the earlier operation for various reasons. Although there are copious amounts of coal, there is still a big problem that companies are forced to face after the long process of obtaining permission to extract is granted. The layers at many of the sites in Britain are thin and often separated by many metres of overburden making coal excavating a job for some of Europe's largest modern-day earthmovers.

England has always been a playground for some seriously large hydraulic excavators. In 1972, O&K delivered Britain's first RH60 – a 100 tonne hydraulic excavator with a 6.1 cu m bucket. This massive machine was soon followed by others that were considered giants at the time, including: the Poclain EC1000 and 1000CK, the O&K RH75, and the Demag H 111.

The 1980s saw bigger excavators – like the Demag H 185, the P&H 1200, the Liebherr R 994 and the O&K RH75C and 90C – arrive in Britain. At the beginning of the 1990s even larger machines, such as the O&K RH200 and Hitachi EX3500, were introduced to British coal mines. However, all these machines were eclipsed by two industry-leading excavators that, at the time, were not just the largest in Britain.

6.6. Over the past few decades, the O&K RH120C has proved one of the most popular mining excavators in Britain. Even after 60,000 to 70,000 hours, many of the survivors are still going.

In 1980, the first O&K RH300 in the world was delivered to Northern Strip Mining for use at its Donnington Extension. At the time it was the very first hydraulic excavator to weigh nearly 400 tonnes, a powerful machine equipped with two Cummins engines that each pumped out 1,210 hp. After only a year at the Donnington Extension, the diesel-powered front shovel excavator was shipped to Godkine coal mine in Derbyshire.

The industry-leading role of the RH300 was surpassed in 1986 following the creation of a Demag H 485 prototype that was commissioned for Coal Contractors for use at their Roughcastle coal mine in Scotland. The diesel-powered front shovel excavator not only broke the 500 tonne weight barrier, it also carried a phenomenal 23 cu m bucket, which was capable of filling a Cat 789 in just four passes.

While the RH300 was scrapped long ago and the H 485 was eventually shipped off to the Canadian Tar Sands, it is estimated that there are as many as seventy active hydraulic excavators, weighing 250 tonnes or more,

in Britain today. The vast majority wear the O&K name because the RH120C was and still is one of the most popular hydraulic excavators in the country: nearly ninety of these formidable excavators followed after the prototype in 1984.

During twenty years of operation, many of the estimated fifty survivors have notched up well over 60,000 hours of active duty: compelling evidence of the machine's solid build. Britain's biggest coal producer, UK Coal, relies on one of the best-preserved fleets of ageing O&K excavators in the country. A number of classic examples of these machines can be found at Steadsburn, the UK Coal operation close to Newcastle in Northumberland.

This site is not only the largest of UK Coal's five active surface mines but also one of the biggest coal mining operations in the UK today. The Steadsburn operation recovered a million tonnes of coal and around 200,000 tonnes of high-quality fireclay over a four-year period. With reserves totalling over a million tonnes, Steadsburn is due to run until 2012; by the time it comes to a close the site will have yielded around 11 million tonnes of coal.

The site produces approximately 5,000 tonnes a week, which means the operation plays an important role in UK Coal's production targets. In 2010 their target was 1.6 million tonnes for surface coal mining.

The fifteen coal seams at Steadsburn vary in thickness and can measure anything from 15 cm to 1 m. With an overburden to coal ratio of 30:1, almost 9 million tonnes of waste rock have to be excavated in order to get at their yearly coal quota. That is a huge amount of rock requiring an average of two blasts a day. On top of that, another 20 million tonnes of overburden must be removed before the mine closes, which means there is plenty of work for the drilling team. Most of the coal is actually buried beneath a thick layer of glacial drift and many metres of bedrock that the excavators must peel off in two layers in order to extract the coal.

The four O&K RH120Cs that work at Steadsburn are around twenty years old. A few of them used to work at Orgreave – a massive reclamation project near Rotherham that spanned ten years from 1995–2005 and produced nearly 5 million tonnes of coal. While most of these machines are still going strong and continuing to pile on the hours, it is a costly exercise to keep these ageing dinosaurs running. Hydraulic pumps and hoses as well as metal fatigue are the major causes of downtime but workshop manager Peter Beattie is full of praise for his ageing fleet. 'To be fair they do not give us too much trouble,' he says.

The excavator's popularity has much to do with a novel technique developed by O&K to ensure fully filled buckets in even the toughest conditions. The problem was a common one throughout excavators: when the teeth of a face-shovel bucket dig into a bench the forces pushing back can be enormous and can push the excavator backwards. The solution that O&K pioneered – called TriPower – was to fix hydraulic cylinders to a pivoting knuckle arrangement on the boom. As the

6.7. It may be old but the RH120C's 13 cu m bucket still fills its Cat 777 trucks in just four passes.

bucket is raised, the weight is transferred to a rear tensioning bar through the knuckle, which exerts pressure on the front of the machine, pushing the tracks to the ground, resulting in a full bucket.

When Terex got its hands on O&K one of the first things the company did was upgrade this popular excavator, the result of which was known as the Terex RH120E. After Bucyrus acquired Terex Mining, both the Terex and the O&K names disappeared completely from this excavator's name. It remains to be seen what further changes are in store should Cat's bid for Bucyrus be approved. (See Postscript.) In the meantime, the mammoth task at Steadsburn is too large for four RH120Cs and a single RH40, so they are assisted by a couple of other massive earthmovers.

O&K RH170

One of the massive machines helping the O&K excavators is a rather interesting O&K RH170 which was launched in 1995 in order to fill the gap between the RH120C and RH200. UK Coal has the prototype, which, even today, is the only RH170 in the country.

Commissioned in South Wales in April 1995, it then went to Orgreave after which it was dismantled and brought to Maidens Hall where it did another twelve-year stint. Incredibly, after sixteen years of duty and over 40,000 hours, it is still going strong.

6.9. After more than 40,000 hours of active service, the machine is still going strong. Its power comes from two Cummins engines which produce a maximum output of 1,686 hp and can burn 230 litres of diesel an hour.

6.8. The O&K RH170 was developed to fill the gap between the RH120C and RH200. The one pictured is the only one in the UK and also the prototype from 1995.

The power of this machine comes from two Cummins engines which kick out a maximum output of 1,686 hp. Burning almost 230 litres of diesel an hour, it is fortunate that the 6,300 litre fuel tank is plenty big enough for twenty-four hours of operation. The O&K RH170 had a couple variations on bucket size: the 345 tonne backhoe version of this excavator carried a 20 cu m bucket, while the 340 tonne front shovel had an 18 cu m bucket. The excavator owned by UK Coal is coupled to a bigger 20 cu m bucket, which has no problem filling a huge Cat 789 truck in just four or five passes.

Although the RH170 was particularly popular in the backhoe configuration in Australia, the front shovel version was never that popular possibly because of the small bucket size. Competitors in the weight class of 350 to 400 tonnes were capable of offering buckets measuring 21 and 22 cu m, while the bucket attachment for the RH170 front shovel remained at 18 cu m for many years. In August 2008, however, things changed with the launch of the new RH170B, which weighed 397 tonnes. After being fitted with a new, stronger, steel-structured design for the boom and stick, both front shovel and backhoe versions of this excavator were able to carry a bucket measuring 22 cu m.

Another major difference from the outgoing machines is that in addition to the Cummins KTA 38 engines, customers can also opt for a Caterpillar C32 power source, while an electro-hydraulic version is also in the pipeline. Impressive as it may be, the RH170 is not the largest excavator at Steadsburn. This honour goes to an even bigger O&K excavator.

O&K RH200

A couple of decades ago, UK Coal boasted the largest fleet of earthmovers in Europe. Although this is no longer true, the Steadsburn site still requires the presence of an RH200 – the largest O&K excavator in the UK – to shift the large volumes of overburden over long distances.

Only four of these excavators were brought into the country: the first two were purchased by R J Budge (later RJB Mining) in March and November 1989 and one year later, two more were added by Crouch Mining.

6.10. UK Coal's O&K RH200 is the last survivor of four that were brought to Britain.
This is one of a pair of machines originally bought by R J Budge in 1989.

6.11. Weighing over 400 tonnes, the huge O&K RH200 carries a 21 cu m bucket, which scoops up around 40 tonnes at a time.

These latter two are no longer in the UK; instead, they are believed to be in Hong Kong. The machines owned by R J Budge were eventually acquired by UK Coal. Interestingly, the excavator that was originally purchased in November 1989 marked the seventh RH200 to leave the production line at the O&K factory.

This massive machine spent much of its life at Orgreave but when the job was finished – after 45,000 hours of labour – it was taken out of active service.

Management considered selling the 1989 RH200 but decided it would be better used for spare parts. Today, the stripped carcass behind the workshop at Steadsburn is all that is left: the machine's vital organs were plundered to keep its sister excavator operational. The choice to use it for parts is clearly paying dividends: using the scrapped machine to keep the twenty-year-old excavator running is highly affordable. So far the slew ring, which had only done 12,000 hours, the hydraulic

6.12. Pictured is the first of the R J Budge machines. After 45,000 hours, it was taken out of active service and UK Coal decided to plunder it for parts to keep the only remaining machine running for longer.

6.13. The RH200 loads a Cat 789, the largest Cat dump truck in the country. With a payload of 177 tonnes, there are believed to be only ten of these monster haulers in Britain; Steadsburn has eight.

pumps and the undercarriage, as well as a number of other components, have been taken from the dismantled machine.

With an operating weight of over 400 tonnes, the only active RH200 that still exists in the UK was actually one of the first to leave the production line after the prototype debuted at the Bauma show in 1989. It was commissioned by West Chevington, the first coal pit at Maidens Hall, and has always remained at that site. During the early part of its life, the sole surviving O&K RH200 in the UK used to operate twenty-four hours a day; a fact that is evidenced by the high number of hours – over 60,000 – that the machine has clocked. Even today it is still expected to be available for two 8-hour shifts per day and at least 3,000 hours a year.

Although the RH200 has clearly seen better days, there is no reason why it should not survive for another five years and clock up an additional 10,000 to 20,000 hours, providing that there are no seriously expensive

breakdowns. The power of this machine comes from two water-cooled Cummins KTA 38 engines which produce a combined maximum output of 2,030 hp. It is not surprising, therefore, that when working flat out the engines do not take long to drain the 10,250 litre diesel tank. Though the original model was fitted with a 20 cu m bucket, a former owner replaced it with a 21 cu m version. Now capable of scooping nearly 40 tonnes at a time, the RH200 can fill the Cat 789 trucks in just four good passes. In tough rock, however, it can take as many as five or six.

The excavator spends most of its operating hours loading previously extracted glacial drift. Approximately 3 million cu m of this material has to be shifted. The mine is covered by the drift, which can be 30 m deep in places, and it is tricky stuff. Although it develops a hard crust on the surface, underneath it never dries, which means that it often does not support the weight of the big trucks and excavators.

6.14. This Cat D10R is the largest of nine Cat dozers at Steadsburn. When launched in 1977,
this Cat was the most powerful dozer in the world with a 700 hp engine.

Cat 789

The original operation, of which Steadsburn is a spin-off, was known as Maidens Hall and located in much the same place that Steadsburn is now. Maidens Hall was a big coal-mining operation that first started in 1980 at West Chevington. The extraction process should have been one big job but, for various reasons, it ended up as four smaller ones of which Steadsburn is the last.

The northern extension of Maidens Hall is now exhausted of coal and quite a few of the unusually large fleet of forty-five trucks are kept fully occupied carrying glacial drifts to backfill the void. Approximately 2 million tonnes of coal were removed from this site resulting in a void with a depth of over 100 m. Now that the operation is shutting down, nearly 21 million cu m of material will be needed to refill the void. Backfilling takes careful planning because if all the softer material is put into the hole at once, then it will not support the weight of the trucks. The solution to this problem is for the trucks to lay down alternate layers of rock and glacial drift.

The job is overseen by a Cat D10R, the largest of nine Cat bulldozers on site. In the early 1970s, the D9 was Caterpillar's biggest dozer but just a few years later, both

Fiat-Allis and Komatsu brought out even larger ones. Cat responded to the competition by launching the 700 hp D10, which would turn out to be the most powerful dozer in the world at that time. Available for purchase in the autumn of 1977, the D10 was a hugely successful machine and very well-liked by British coal mines.

The fleet of trucks helping to fill the void at Maidens Hall is comprised of twenty-seven Cat 777B, 777C and 777D versions. Many of the older machines that were inherited from Crouch have clocked up anything from 60,000 to 70,000 hours of operation. The twenty-seven Cat trucks, along with three Terex TR100s, are all used to transport material excavated by the O&K RH120s. However, all these 91 tonne capacity trucks are dwarfed by a number of monster Cat haulers. The Cat 789 is the largest truck in the UK, with a payload of 177 tonnes. There are only ten of these massive trucks in the UK and Steadsburn has eight of them. These particular trucks were purchased in 1989 by UK Coal, but they were not the first ones in Britain. In fact, Scotland was the first place to have the Cat 789 but these have long gone, as is the Demag H 485 that used to load them.

6.15. Although UK Coal runs the original A series 789s, they have been upgraded to B series specifications.
When the 35 litre engines needed replacing, they were swapped for electronic ones.

The long haul distance means that seven of the Cat 789 trucks are needed to keep the RH200 working to full capacity. However, it is unlikely we will ever see many more of these Cat trucks in the UK in the future. Today they are simply too large for most British mines whose preference is for a three-pass fill of a 90 tonne truck.

While UK Coal originally bought the A series 789, they have since upgraded them to B series specifications. When the 35 litre engines needed replacing, they were swapped for electronically governed ones. While there were a few other tweaks, this engine swap was the main change between the A and B series machines. Just like the big excavators at the mine, the oil coolers and engine radiators on the 789s are fitted with noise deflectors that were developed by a local engineering company to ensure that as much sound as possible is diverted to the ground.

On a good day, all eight trucks can be relied upon and

workshop mechanics often joke that the trucks have only just been broken in. The truth, however, is that these hauling dinosaurs need constant attention: after twenty years of service and over 60,000 hours of operation, they need patching to keep them going. Metal fatigue is one of the biggest problems.

Case in point, the Chevington Collier walking dragline used to keep four welders fully employed with two of them spending six months on refurbishing just one of the massive buckets. When the dragline was decommissioned, these mechanics did not lose any work because they had to channel their energy into keeping the ageing fleet of big trucks and excavators operational.

Welding is not the only task carried out at the workshop. Mechanics also stock reconditioned radiators and engines there. These component parts for the 789 are not cheap: a re-built engine costs around £100,000. Despite the stocked parts and the fact that many of the big trucks are approaching 70,000 hours of operation,

6.16. Dust is a sensitive issue at any mine. Three converted Cat 777B trucks help Steadsburn keep on top of it.

mechanics are confident that all eight will still be going when the job at Steadsburn is finished.

If the mechanics' predictions are correct, then UK Coal will have to decide what to do with the monstrous trucks. One option will of course be to transfer them to another site but this will not be an easy solution. Removing the truck bodies in order to transport them on their own is not enough. Transport restrictions state that they need to be cut vertically into two pieces, a process which would entail two or three welders taking a couple of days to slice through the 7 m wide skip.

It is not only the future of the trucks that will soon need to be decided, but also what to do with all the big machines when Steadsburn closes in 2012. There are a number of options for the RH200: it could be sold as an active machine alongside the pickings from the carcass of the first one or it could be sold to countries – such as those in Asia – where the labour rate to repair is much cheaper. Alternatively, the RH200 might be worth quite a bit if scrapped. One final option is that it may be transferred to another surface mine in the UK.

Peter Beattie is convinced that the big excavator still has several years of life left in it yet. 'Providing there are no seriously expensive issues, there is no reason why it should not be able to clock up 70,000 to 80,000 hours of operation.' Unfortunately, if there is a major problem with a part that cannot be stolen from the scrapped machine, then it is quite likely that UK Coal will not pay for repair. At the moment, the six O&K RH dinosaur excavators are a great public relations exercise for both UK Coal and Bucyrus; however, it is unlikely that any of the old excavators and trucks will still be around in 2020.

Terex O&K RH200

The immense physical size, weight and age of the O&K RH200 puts it in a league of its own, but, while it is the only remaining excavator of its kind in Britain, there is also a newer Terex O&K RH200 model left in the UK. The Terex excavator was purchased by Banks Mining and started work at the Shotton surface mine in Northumberland, which is a relatively new site that only opened in May 2008.

The Shotton site has been divided into three phases. Phase one, which contains just two coal seams, is currently operational. The good news for Banks Mining is that it has permission to extract nearly 5.4 million tonnes of coal from phase one, two and three. The bad news is that they have until 2018 to perform the whole operation, including the reclamation work. With heaps of overburden to remove, the short time-span is a tall order. Although the first coal layer is only 5 m below the surface, the deepest layer is around 100 m.

In excess of 200,000 cu m of overburden will need to be extracted in order to excavate just 15,000 tonnes of coal each week. Almost 2,000 tonnes of this coal is burnt in British homes during the winter. Much of the rest goes to Blyth Docks for rail transport to Drax power station in North Yorkshire. There is also demand from the coal-fired power station located at the local aluminium smelting works, owned by Rio Tinto Alcan.

At this mine, overburden extraction is the job of four large excavators. Two of the four are relatively new E series 120s and, of the twenty excavators of this type in Britain, the ones at Shotton are also the newest. The RH120E has undergone a number of minor changes over the years. While the first ones had an operating weight of 265 tonnes, the latest ones weigh considerably more – over 280 tonnes.

The older RH120C was traditionally offered with a pair of Cummins KTA19 engines. However, the two RH120E excavators at Shotton are powered by Cat C18 engines. Two of the Caterpillar C18 engines fitted in one of the RH120E shovels were prototype units being tested by Terex and Caterpillar to determine if they could overcome certain problems they had encountered with the original C18 option offered in the RH120E.

Following the test period, Terex and Caterpillar offered to replace the Cat C18 engines with the larger Cat C27 ones in order to address the issues. Caterpillar claims that the C27 engine will be more fuel efficient than the C18 in this particular application. The C27 engine is more powerful and responsive and the company claims that engine life should be much longer.

This extra power and weight allows the RH120E to carry a larger 15 cu m bucket but, even with the help of one older model O&K RH120C, the combined efforts of the three excavators cannot shift enough overburden to expose the weekly coal requirements at Shotton. The solution to this problem is an eleven-year-old Terex O&K RH200, a popular excavator because of its durability and endurance. So far, the combined efforts of over a hundred of these huge excavators, since the launch of the very first one in 1989, have clocked up over 3 million hours of operation.

6.17. Two of the newer Terex O&K RH120Es earn their keep at Shotton. Commissioned in the autumn of 2008, these were the first two machines in the country to be fitted with Caterpillar 18-litre C18 engines.

6.18. At 525 tonnes, the Terex O&K RH200 is quite easily the biggest hydraulic excavator in England today.
The excavator carries a 26 cu m bucket capable of loading 50 tonnes at a time.

6.19. Equally impressive are the two Cummins engine options. The machine owned by Banks uses the 38-litre Cummins KTA 1200
which produces a maximum output of 2,400 hp at 2,100 rpm.

The Banks Mining excavator was built after the Terex Corporation acquired the O&K mining division in 1998 – hence the name Terex O&K RH200. The massive excavator was brought to Shotton from another local mine. Terex engineers dismantled, moved and reconstructed the machine: just ten days later it was back at work.

The Terex O&K RH200 weighs 525 tonnes, making it not only the largest hydraulic excavator in Britain today but also the one with the biggest front shovel bucket. While the first O&K RH200 carried a mere 20 cu m bucket, the Terex O&K RH200 supports a massive 26 cu m version, which is easily capable of handling 50 tonnes of material at a time.

Equally impressive are the two Cummins engines available for the Terex model. The machine owned by Banks Mining uses the 38 litre Cummins KTA 1200 which pumps out a maximum output of 2,400 hp at 2,100 rpm. Also impressive is the 10,250 litre tank, which has to be topped up every day for the excavator to run but would fill the average family car almost 150 times as well as keep it running for around two years.

Less Noise

Noise reduction is a crucial issue for most surface coal mines and the thirty-five in Britain are not exempt. Not so long ago excavator engines and hydraulic coolers could be heard for many miles but times have changed. Today it is vital to contain the sound of working machines to the mine.

Shotton is certainly more noise sensitive than most sites and Banks Mining needed to satisfy a couple of important criteria before they could obtain a coal extraction licence for the site. The first requirement was to create an 8 m high bund on one side of the site in order to shield it from a residential area. To further reduce noise levels the company added extra sound damping to the engines of the big trucks. The excavators were already fitted with a number of noise reduction modifications to insulate the engine compartments and splitter silencers to quiet the radiators and hydraulic oil coolers. New exhaust silencers have been designed, fitted with tightly packed mineral wool, which also absorbs noise. The result is that noise levels on the Terex O&K RH120 have been reduced from 90 decibels to 80 decibels.

By adding secondary silencers to those already fixed to the exhaust of the larger Terex O&K RH200, even more impressive noise reductions can be reached. Noise levels are measured 10 m from the machine with the engines running at full power: after the secondary silencers were fitted, the noise levels fell from 97 to 82 decibels! Banks Mining is a pioneer of noise-reduction technology and now their machines are attracting attention from some of the biggest mines in the world that are keen to follow suit. Having come up with a solution to the problem of noisy excavators, the next problem for the team at Shotton to tackle will be the noise emissions from the tracks of the bulldozers.

Dust Prevention

Machine noise is not the only issue faced by Shotton. The site is flanked by two pharmaceutical companies, which means that there are strict policies on dust. The brief from officials to Banks Mining was simple: comply with regulations regarding dust or planning permission to develop the site will not be granted. So Shotton employed a number of measures to control dust, including: three water bowsers to cover the haul roads and an automatic sprinkler system to cover the coal stockpile.

However, the biggest potential source of dust at any surface mine is created when excavator buckets empty their contents into a truck. After looking hard at ways to address this, they developed a novel dust-suppression system for each of the prime movers. The upper structures of all four prime movers feature a series of massive tanks, which supply water to two spray lines. The lines have four nozzles each and are located above and below the front shovel. Spray is automatically activated when the crowd arms are retracted.

From a distance it looks like a dust haze at the top and bottom of the shovel bucket but it is in fact a fine mist of water. The theory is that dust particles will attach to the water droplets and fall to the ground. The clever system plays a vital role in keeping dust levels to a minimum.

Prototyping work was done in-house on a RH120C. Thereafter, Terex engineers came to Shotton to assess the new system, before returning to Germany to design the best way to incorporate this into a brand-new excavator. The resulting design is a strengthened upper carriage to support the extra weight of the water tanks on both of the new RH120E models. There are three water tanks in total: two on each side and one at the rear. They give the machine a physically larger appearance and are capable of carrying 10,000 litres of water when full.

The Terex O&K RH200 has also been fitted with the dust-repression system. Unlike the RH120E models, this excavator has four massive water tanks in total. Capable of holding 16,000 litres of water combined, they still need a top-up after just four hours.

6.20. The spray nozzles are located on the stick above the shovel bucket of the RH120C.
They quickly suppress any dust that escapes when the bucket teases material out of the ground.

6.21. This twelve-year-old O&K RH120C is the oldest of three 120s at Shotton. It looks a little different to
all other 120Cs because of the side-mounted water tanks for the dust suppression system.

6.22. The Terex O&K RH200 at Shotton is fitted with massive water tanks at each corner of the upper structure. The giant excavator easily fills each of their five 136 tonne capacity Cat 785s in three attempts.

6.23. The 85 tonne Cat 385 is a big excavator that was originally born out of the 5090B – an older 5000 series machine. The machine at Shotton spends most of its time removing interburden between thin layers of coal.

Big Cat in the Yorkshire Dales

The story of the Caterpillar 5130B hydraulic excavator was well-covered in Chapter 5 (Spain). However, one of these machines works in England, in the middle of the Yorkshire Dales National park, perhaps one of the most unlikely places for an excavator of its size, and thus it requires another look.

Only four of these excavators ever came to Britain and one was also the very first 5130B in Europe. The excavator was purchased by Wimpey Mining for use at the Ffos Las coal mine in Wales – at the time the largest surface coal-mining operation in Europe. At the site, the 5130B was responsible for loading Caterpillar 777 and 785B trucks. Eventually, the first and second 5130B excavators were shipped out of the country, while the third – believed to be a unique cross between the original ME and the newer B versions – stands idle in between coal-mining jobs in Scotland.

The 5130B is the largest working Caterpillar excavator in the country. The machine is the prime mover at the Swinden quarry, near Skipton, also known as the second largest Tarmac quarry in the country. Of the 2 million tonnes of aggregates extracted here, some is used for dry stone walling and the rest is used internally to make concrete or asphalt. Almost fifty per cent of production is transported from the site by rail: interestingly, the site employs the heaviest locomotive in Britain to track out the nine trains that leave the site every week.

There are no big trucks in this quarry. Instead, the excavator is positioned on a pad, a lofty 4 m above the ground, and with the added height it is able to load one

6.24. At one time there were four Cat 5130Bs in Britain. This example is working in the middle of the Yorkshire Dales National Park; it is the only active 5130B in Britain today.

of the largest mobile crushers in Europe – a 240 tonne monster that is capable of devouring limestone blocks measuring 1 m in diameter. The volume of material put through each year is equivalent to laying 1 m square blocks end-to-end from London to the tip of Scotland.

The two larger-than-average machines were bought in 2000 as part of a £16 million investment to reduce the quarry's impact on the environment. While their two former excavators and surrounding truck fleet used around 4,000 litres of diesel a day, the new arrangement has not only halved the fuel requirements of the quarry, it has also slashed the wage bill because twelve operators are no longer needed.

The excavator shares a number of parts with other well-known Cat machines. For example, the 34.5 litre, 800 hp engine also provided power for the 777D truck, while much of the track drive system comes from the D11 dozer. However, this excavator differs from other 5130B models in that it carries a longer stick – measuring 5.2 m – which replaces the standard 3.9 m version. The change to the stick was for good reason: the extra metre of reach, which makes the total reach just over 18 m, is a big bonus because operators need to be able to handle as much rock as possible from a single position. It takes time to move the crusher, time that leads to lost production.

The downside to the longer stick is that the original bucket, which measured 10 cu m, had to be swapped for a smaller, 8 cu m one. Even with this change, the excavator can still comfortably lift 15 to 20 tonnes of material at a time, filling the hopper of the crusher with 9,000 tonnes a day.

For the past ten years, the excavator has done nearly 26,000 hours and has given quarry management little trouble. During the past three years the reliable workhorse lost an average of just four production days a year through breakdowns – not bad for a machine that does 3,000 hours of operation a year.

At Swinden quarry, there is another 24 million tonnes of limestone to extract before 2030 when the licence is due to expire. Although the crusher will outlive the project, quarry management recognise the need to decide the future of their excavator. One option would be to keep it until it has clocked 30,000 hours and then give it another engine and hydraulic pump update. Or they could part-exchange it for a brand-new excavator – though not for another 5130B because production of that model ceased in 2003.

Although the modified 5130B still has a few years left, management are already looking at all available options. When the time does come to sell the massive machine there is a good chance that the 185 tonne excavator will be exported, meaning that in a few more years, the 5130B will be banished to the British, and even European, archives.

Cat D11R

For the past thirty-five years, Stokey Plant Hire has built a solid reputation as the place to go to hire a big dozer, such as a Cat D9 and D10. The company truly rocked the British earthmoving world when it purchased a D11, weighing in at 110 tonnes. This machine, the only one in the country, was bought for a five-year contract at Tarmac's Croxden aggregates quarry, near Cheadle. The massive dozer was charged with the task of ripping and bulldozing one million tonnes of material a year from the site, which is believed to be Europe's largest deposit of sand and gravel. The deposit was laid down about 240 million years ago and in places the material is nearly 50 m deep.

The massive machine was bought in as a replacement for another big Cat dozer – a D10. Some of the rock on the Croxden site was so tough that the D10 really struggled, so Stokey Plant Hire decided that more power was needed and bought the bigger, 850 hp D11. This machine was the very first one in the UK and it spent many hours ripping and pushing.

The conditions at the quarry are so tough that even Caterpillar's biggest dozer was forced to knuckle down; the wear line on the heavy duty ripper, counterbalanced by a larger blade at the front, clearly confirms that this machine spent a lot of time at full depth. Ripper wear parts are subjected to serious abuse at the site and it is not uncommon for the machine to use up to seven teeth a day on some of the bottom benches.

While quarry management and operators alike have had no qualms about the performance of the machine, operator visibility, especially to the rear, was an issue, and a couple of safety measures had to be taken to improve it. The rear corners of any bulldozer are notorious blind spots. On one side the fuel tank obscures visibility, on the other side it is the hydraulic tank, while the roll-over frame also impedes the view from the operator's seat.

To improve things at the rear, three external cameras were installed – one on the ripper and one on each side

6.25. With an operating weight of over 100 tonnes, the D11R is the largest Cat dozer in the country. Stokey Plant Hire has the only one in England.

of the tracks. At the front, convex mirrors were put in place to allow the operator to see the tracks and whether there was anyone on them.

Supervisor Chris Chamberlain sat on the D11 for the first time in 2009 and he admits that initially he found it a real challenge. 'In terms of size and power it is in a different league to the D10,' he says. 'I love it. It is a great machine to drive and it gives me a real buzz.' However, he confesses that it can be a rough ride. 'Ripping and prising out rocks that are sometimes half the size of a Mini Cooper results in lots of jolts and jerks. Afterwards, you still have to get back over them to be able to push them out of the way.'

Sadly, after six years of reliable service, the future of the only D11 in Britain was uncertain. In 2009, the management at Swinden sought planning approval to extend the operation. When the recession hit, these plans changed drastically. The demand for aggregate slumped and Tarmac decided to close the Swinden operation and reduce stocks until things improved. The D11 was steam-cleaned and Stokey Plant Hire searched for a buyer with the most likely outcome being that the bulldozer would disappear from British shores.

Fortunately, after standing idle for the best part of a

year, the huge dozer was recently taken off the market and cranked back into action at the newly reopened Croxden quarry. At least for the next few years England will once again be the home of the most powerful Caterpillar dozer in the business.

Komatsu WA900

Over the years, the Ketton limestone quarry, owned by Hanson Cement, has received its fair share of publicity. In April 1972, the quarry, which is located just across the other side of the A1 from Stamford and is home to a 63 m long bat cave, shot to fame following the discovery of various Bronze Age relics. The items uncovered are known as the Ketton Hoard and include sixteen axes and a broken dagger; they are on display in Oakham Museum.

Then there was the mystery of the 'black cat', thought by some to be a panther, seen in the quarry: the story was featured in a BBC television documentary. Finally, the quarry was selected as the location for a Top Gear programme in which presenters spectacularly shot cars off the quarry ridge onto a dartboard painted on the floor below!

6.26. The WA900 is the most powerful Komatsu wheeled loader in England.
The Ketton quarry – owned by Hanson Cement – has the only two in the country.

Despite recent events, the quarry's main function is not to accommodate television crews but to yield limestone and clay to make upwards of 2 million tonnes of cement a year. The material from this quarry is used countrywide. High-profile successes include supplying cement for the second Severn Crossing, the Jubilee Line tube extension in London, the Thurrock Viaduct section of the Channel Tunnel Rail Link and the Medway Bridge M2 motorway-widening project in Kent.

The 85 hectare quarry hides an intriguing array of different rock layers. The layer of high-calcium limestone they are after is 11 m thick and was formed more than 90 million years ago, during the mid-Jurassic period. It is concealed by topsoil and a layer of Blisworth limestone which is 3 m deep. The Blisworth limestone, as well as the topsoil will eventually be used in the reclamation process. Then there is another 8 to 10 m of clay, the majority of which is used in the cement-making

process, and finally the layer of limestone. Also used by the quarry is the 2 m layer of silica limestone below the high-calcium limestone layer, as is an even deeper layer of ironstone.

The only two Komatsu WA900 wheeled loaders in England work at the Ketton quarry; with hourly targets of 2,000 tonnes of cement to produce, there is plenty of work for the prime movers. Although well-liked in the US, this popularity has yet to be repeated this side of the Atlantic. In addition to the two machines in England, there are just a handful of these machines in mainland Europe; consequently, not that much is known about the WA900.

The pair at Ketton quarry has worked there since 2007, where they are each clocking up between 2,500 and 3,000 hours of operation a year. They are slightly bigger and heavier than the two Cat 992G wheeled loaders they replaced, and neither the quarry nor Komatsu has made much noise about the newcomers.

6.27. The giant loader has a 13 cu m bucket that helps the machine to hit peak production figures of over 1,200 tonnes per hour.

This may be because the WA900 shares so many similar vital statistics and specifications to the WA800. The WA900 does, however, feature more weight and extra muscle power, hence a bit more productivity, which was something the Ketton production team were keen to put to the test.

The WA800 weighs exactly 104.5 tonnes, while the operating weight of the WA900 tops that by nearly 3 tonnes – mainly by adding more counterweight to the rear bumper. Although both machines share the same 30.5 litre V12 engine, the WA900 manages to squeeze a bit more power out of it. The newer model can manage 901 hp at 2,050 rpm whereas the WA800 is capable of just 853 hp at 2,000 rpm.

The extra weight and power may not sound like much, but they are sufficient to allow the WA900 to wield a 13 cu m bucket whereas the WA800 can only manage 11 cu m. The extra 2 cu m makes all the difference and at peak production the WA900 shifts over 1,200 tonnes per hour. The two massive wheeled loaders fill a fleet of six Komatsu trucks – three HD605 models, each with a 63 tonne capacity, and three HD785 models, with a 91 tonne capacity. The HD785 trucks use same V12 engine as the massive WA900 wheeled loader.

Capable of scooping up 20 tonnes at a time when digging in limestone or as much as 25 tonnes when digging in heavy clay, these machines require just three passes to fill one of the 63 tonne trucks. Depending on material, four to five attempts are usually needed to send a 91 tonne truck on its way. Management is happy with the rate of work produced by the WA900s since there is plenty of work left at Ketton for these massive machines. The limestone reserves are estimated at 50 to 60 million tonnes, easily sufficient for the next twenty years.

All cement produced Ketton is a mix of eighty per cent limestone and twenty per cent clay. While there is clearly no shortage of limestone, the reserves of clay are not as dependable. Clay is a vital ingredient in cement. A fully heaped WA900 bucket of clay is added to every truck before it heads off to the crusher. Once at the plant, the material is heated to 1,300 degrees Celsius in the kiln and a chemical reaction takes place. The resulting clinker is ground into the powder to make up what we know as cement.

Despite the abrasive conditions of the quarry there are no chains on the tyres of either wheeled loader. Instead, the quarry prefers to weld deflector plates to the sides of the bucket to divert rocks away from vulnerable tyres. Operator John Jamieson – one of four drivers trained to control the big loaders – is impressed with the joystick steering system, the good all-round visibility and the cab ergonomics, such as the elbow rests. 'The design of the WA900 has been well thought out, obviously with operator comfort in mind,' he says.

CHAPTER 7

Luxembourg

When you hear about the country of Luxembourg – landlocked by France, Belgium and Germany – a few interesting facts should come to mind. One is that it is the world's only remaining Grand Duchy – officially called the Grand Duchy of Luxembourg. Second, fuel prices are the cheapest in Europe and motorists flock in hoards to replenish their tanks, patiently braving queues of up to an hour. Third, Luxembourg is one of Europe's smallest countries, measuring just 84 km long and 52 km wide. Lastly, with a population of roughly 500,000 people, Luxembourg is also one of Europe's most thinly populated countries. With all these quick facts to consider is it any wonder that many people do not know that Luxembourg is also home to a machine that will get the heartbeat of any hardcore machine fanatic racing? That machine is Caterpillar's biggest tracked dozer!

Caterpillar claims that they have made more than 3,000 D11s at the East Peoria facility in Illinois and although the majority of those work in the US, the bulldozers are also popular in Canada, South Africa and Australia. The reason they are not so popular or common in Europe is that European mines and quarries are simply not big enough to warrant one of these huge machines. During the past ten years, only forty-two D11s have made it across the Atlantic so it might come as a surprise to learn that one of these massive 105 tonne machines resides in the tiny country of Luxembourg. The question is why is it there?

7.1. With an operating weight of 105 tonnes, the D11 is the biggest tracked dozer that Caterpillar produces. This is one of seven new T series machines in Europe, and the only one in Luxembourg.

The answer lies with Luxembourg's history of steel-making and the slag dumps that were an unwanted by-product of the industry. Steel-making in Luxembourg has been an active tradition for over 150 years and many of the slag dumps have remained undisturbed for over a century. ArcelorMittal, the largest private employer and steel maker in Luxembourg, recently embarked upon the task of removing them. The company has contracted Cloos, Luxembourg's largest aggregates and recycling company, to process all current slag from its three electric-arc steel mills and clean up several older dumps left by former blast furnaces. Cloos began handling slag in 1902, so it has immense experience making it a true expert in the field.

In their hundred years of operation, Cloos has employed a variety of methods for extracting and recycling the slag. The days of removing the slag by hand are long gone, and Cloos now relies on the work of heavy-duty machinery, such as the 65 tonne Caterpillar D10T dozer. Until recently the company had three of them – one at each of three big slag heaps – but now that those sites have been cleaned, all the attention has been diverted to the last major blast-furnace slag dump in the country.

This massive heap of slag, called Terres-Rouges (Red Land), is located 20 km south-west of the capital at Esch-sur-Alzette. Terres-Rouges is so large that it sprawls across the French border; originally mined by ARBED, which later became ArcelorMittal, the area forms part of a huge underground belt of iron ore that spreads for many miles. The first furnace was fired up on April 10, 1872, and by 1899 three more had been constructed. In 1911, they were all connected to the mill in Esch-Belval, forming a massive steel complex.

However, demand for the product started to fall in the 1950s, and all but two of the furnaces were closed. By the spring of 1977, the plant stopped all steel production and the blast furnaces were demolished. Today, the massive slag heap is all that remains of the enormous site that was once Terres-Rouges. Measuring 25 m high in some places, there are no accurate figures as to just how much slag it contains. The most educated estimate places the total at somewhere between 20 and 30 million tonnes.

Processed slag is a major component in the production of asphalt, which is used to build roads. Despite the fact that there are no major road-building projects on the horizon, the past two winters have wrecked Luxembourg's motorways so much that there is currently a healthy demand for the material to repair them. As a result,

7.2. The Terres-Rouges slag heap is huge, in some places a steep 25 m climb to the top. It should keep the D11 dozer busy for the next ten to fifteen years.

7.3. *Terres-Rouges is one of the biggest slag heaps in Europe. Though there are no records of the amount of slag it contains, estimates range between 20 and 30 million tonnes.*

Cloos removes slag from Terres-Rouges at the rate of nearly 2 million tonnes a year.

The decision to bring in the Caterpillar D11 bulldozer, one of seven T series machines sold to Europe, was made in response to the need to extract and process the final heap of slag in Terres-Rouges. This latest version is a good deal heavier, more powerful and sophisticated than the very first D11N launched in 1986 to replace the D10. The D11N was powered by a 770 hp engine, weighed 92 tonnes and was capable of pushing 34 cu m of material. However, its power was surpassed in 1996 at the MINExpo in Las Vegas where the N series gave way to the heavier

7.4. *The massive hydraulically powered ripper is ideal for loosening the compacted material. The machines must work hard in order to produce 2 million tonnes a year.*

7.5. Generating 850 hp, the D11T oozes power. The C32 engine is barely stretched and pulls the hydraulic ripper at speed in even the toughest conditions.

7.6. The purchase of the D11 allowed Cloos to get rid of two of their D10s. The D11 is truly unstoppable when it starts pushing.

and more powerful R series. This machine had an 850 hp engine and weighed 104 tonnes. Also introduced in 1996 was the D11R CarryDozer (CD) option, which could push 52 cu m of material. These machines were much more refined and could be steered from a single left-hand joystick.

Sophistication levels increased still further with the launch of the D11T during 2008; in this year electrical switches replaced manual levers on most machines. Another addition was rear-view cameras, and new options included hydraulically powered access steps to improve operator safety. Cloos bought their D11T nearly a year ago and it is the ideal machine for tackling the last heap of slag. While their D10T can do ninety-five per cent of the work, the larger machine can pick up the slack when the going gets tough.

The most crucial duty of the D11 dozer at Terres-Rouges is to rip the solidly set slag so that it can begin scraping off the top layer. A quick look at the wear on the paintwork of the rear shank confirms it is regularly used it at depth. When pushing over 40 cu m at a time, aligning the tracks over the ripped area helps to maximise traction. A heavy machine like the D11 is handy for this job because it crushes big lumps of slag as it rides over the top.

The D10 at Terres-Rouges has quite an easy time as it only works a single 8-hour shift. However, the D11's workload is exactly the opposite, since it is expected to do twice the number of hours of the D10.

7.7. Hydraulic rippers do not get much bigger than this one. Blast furnace slag is tough material and the worn paint confirms that it regularly works at maximum depth.

Operator Cyril Brocas used to drive one of their older D10s. Even though he has sat on the D11 for nearly a year he still gets a real buzz from driving it. 'There is no comparison,' he says. 'It is a formidable, virtually unstoppable machine.' His only criticism is rear-view visibility. He says that the machine's fuel tank impedes visibility on one side and the hydraulic oil tank blocks it on the other side. The option of rear-view cameras greatly improves things but it still pays to stay well away from this machine when it is working.

7.8. Working two 8-hour shifts a day, operator Cyril Brocas reckons the experience is hard to beat. 'It is a fantastic machine to drive,' he says.

7.9. Nearly 10.5 m long and 5 m high, the giant D11 dwarfs the fuel truck. The dozer burns over 120 litres of diesel an hour when working flat out, which means the 1,987 litre tank needs filling daily.

Most D11s work in mines and quarries, often ripping up huge rocks half the size of a small car. This work can be rough on the machines as they have to climb back over the rocks before they can push them out of the way. By comparison, the top of the slag heap is quite flat. Even so, the dozer works hard and when the ripper is at maximum depth the engine can burn over 120 litres of diesel an hour.

As the country's largest remaining slag heap, Terres-Rouges could keep their new D11T and older D10 busy for the next ten to fifteen years. Once the slag heap disappears, however, Luxembourg is unlikely ever to see another D11 bulldozer. Even though steel still accounts for a quarter of Luxembourg's export trade, when the Cloos project at Terres-Rouges is complete, the country's D10s are quite big enough to cope with current slag volumes. The fact is that the D11 is not only too big and difficult for many earthmoving contractors to move around – it is also too expensive. Even second-hand ones are not cheap; used D11Ns from 1989 still cost anything from £200,000 to £300,000 and sometimes more!

7.10. The D11T is now available with a hydraulically powered access ladder. This makes it much easier and safer for the operator to climb in and out of the cab.

CHAPTER 8

Switzerland

The impact of the recession on the construction industry has been apparent all over Europe and with fewer builds it comes as no surprise that there has been a sharp fall in the demand for cement, which has, in turn, forced many quarries to lower their production forecasts. Switzerland, however, appears to be weathering the effects of the economic downturn better than most. The demand for cement in the country is at an all-time high, due, in part to the need to complete a number of very large construction projects.

One of the biggest developments is a new 57 km rail link through the Gotthard tunnel in the Swiss Alps. When it opens for service in 2017, the length of this Swiss rail link will not only exceed the 53.8 km Seikan rail tunnel linking the Japanese islands of Honshu and Hokkaido, but also the world's longest road tunnel, the 24.5 km Laerdal in Norway. The future world's longest rail tunnel will shave off an hour of a train journey between Zurich and Milan, making the journey just two and a half hours. Officials estimate that it will be utilised by as many as 300 trains a day.

A true feat of modern-day engineering, the enormous project has cost €9 billion. In places the railway line is 2,000 m below the surface and the whole project has kept 2,500 staff busy for nearly twenty years. To put it in perspective, if the 13 million cu m of rock that had been cut away by the massive drill rig during the past fourteen years was lumped together there would be enough to build five of the world's largest pyramids!

However, this is not the only mammoth construction requiring heaps of concrete in Switzerland. There are also quite a few new roads being built – such as the brand-new Jura Express, a motorway which will run through the rugged north-west of the country.

8.1. The vehicles owned by Jura Cement are a familiar sight on Swiss roads.

8.2. Sixty per cent of production leaves by rail and all is destined for Swiss construction projects.

Swiss limestone quarries are definitely making the most of their ability to satisfy the overwhelming demand for cement. Five million tonnes of cement is produced each year and more than half of that comes from the Jura region. Flanked on one side by France and on the other by Germany, the hillsides of the Jura are dented with numerous smaller quarries, but Jura Cement is one of the three biggest operations.

The headquarters of Jura Cement is based at Wildegg, a short drive from Zurich, where the 125-year-old Oberegg and Unteregg quarry is located. Both Jura Cement and the quarry are owned by the Irish CRH Group. Oberegg and Unteregg is a unique example of a quarry site that is keen to explore all avenues to reduce their effect on the environment. In fact, Switzerland as a whole has built up an enviable reputation not only for efficiency, punctuality and quality, but also when it comes to environmental issues. Given the stunning natural beauty of the country it is perhaps easy to understand why they would want to keep it as green as possible.

Even a decade ago, quarry management researched new ways to reduce their dependence on the fossil fuels that they use to heat the cement ovens on site. Today, sixty-five per cent of their energy requirements come

from alternative means. In addition to bonemeal and several thousand tonnes of solvents, each year they burn 50,000 tonnes of shredded tyres and 50,000 tonnes of plastic, and are about to start turning wood chips into energy to heat their ovens. Although they still need 40,000 tonnes of coal a year, this is a massive reduction on what was previously used.

The limestone that they mine at Oberegg and Unteregg is good quality and extremely plentiful in supply. The seam has a depth of nearly 30 m and the only problem is that in some places it is buried beneath a layer of mainly topsoil, which is almost 20 m deep. With all this deep excavating, the workers at the quarry have uncovered a few surprises: they recently found a 30 tonne boulder, thought to have been deposited during the Ice Age, and a Mammoth tusk!

Ever mindful of their surroundings, the quarry tries to uncover the limestone seam with as little disturbance to the surrounding communities as possible. They prefer to utilise the power of pushing rather than explosives, so the blast sirens see very little action. Since 1985, they have been relying on the largest tracked dozer in the country – the only Cat D10 in Switzerland – to push off the topsoil and expose the limestone.

Since 2009, the ovens have been working non-stop because cement production topped one million tonnes. The combined efforts of the D10 and the smaller D8 are no longer enough to cope with the level of cement outputs that Jura Cement now produces. After considering the purchase of another D10, they eventually opted for a bit more power so they purchased the very first D11T – Caterpillar's biggest tracked dozer – in the country.

With an operating weight of 105 tonnes, the ground really does tremble as this machine roars into action. The D11 simply oozes power: it works for eight hours a day and seems barely stretched by 2,000 hours a year. The combined efforts of the new pushing monsters are already paying dividends with cement production on the rise by ten per cent to 1.1 million tonnes during 2010. Supplied by Swiss dealer Avesco and fitted with track pads measuring 80 cm, the D11T helps to provides a lot more traction and it easily pushes 50 tonnes at a time, while the D10 can 'only' manage 40 to 45 tonnes.

8.3. Each year the quarry gets through a staggering 50,000 tonnes of used tyres, which are used to heat the ovens. This helps to reduce fossil fuel requirements.

The management at Jura is very satisfied with the performance of the D11T and operators think that it is between twenty and thirty per cent more productive than the D10. Another pleasant surprise for the Jura team has been the fuel performance figures of the D11. Although the D11 is much heavier than the D10, new engine technology means fuel consumption is actually on a par with the lighter machine. They reckon the dozer uses around 100 litres per hour when pushing and when ripping – which it does about twenty per cent of the time – fuel consumption rises to a maximum of 110 litres per hour.

While the quarry is pleased with these performance figures, the Swiss government is more interested in eliminating the soot particles and smoke emitted by the exhausts of their machines. Road traffic is not the only contributor to pollution. As far back as 1995, earthmoving machines were thought to be responsible for twenty-five per cent of soot emissions in Switzerland. Alarmed by this figure, the Swiss government introduced new legislation making it compulsory for all new engines over 24 hp and all used engines over 49 hp to be fitted with a one hundred per cent particulate filter system. The only country in the world to do this, the Swiss move has since fuelled a huge market for retro-fit kits.

8.4. Jura Cement uses an old D10.

8.5. The D11T – owned by Jura Cement – is not just the largest Cat dozer in Switzerland but also the only one.

8.6. It is always difficult to appreciate the size of a big earthmoving machine.
However, this Land Rover provides a good idea of the enormous size of the largest Cat dozer.

8.7. With an operating weight of 105 tonnes, this is one of forty-two D11s that have been sold in Europe over the past ten years.

8.8. The combined efforts of the two big dozers helped the quarry to boost cement output by ten per cent to 1.1 million tonnes during 2010.

8.9. The D11T oozes power and is easily capable of pushing 50 tonnes at a time. The D10 can 'only' push 40 to 45 tonnes.

8.10. Exhaust gases from this dozer are re-routed to the cab roof where they are passed through a particulate filter. In Switzerland, this is a compulsory requirement on many diesel engines.

German-company HJS Exhaust Systems is one of several specialists offering approved retro-fit kits to eliminate soot. During the past decade this company has delivered more than 20,000 systems, one of which was fitted by Clean-Life, an HJS partner, to Jura Cement's Cat D11T. The result of this retro-fit on the new dozer does look a little odd; two massive pipes convey exhaust gases to the top of the cab, where they are fed through a strange looking box that houses the diesel particulates filter. Operator visibility is slightly obscured but this a small drawback compared with the cost of the modification to eliminate smoke. Approximately £50,000 was spent on retro-fitting the dozer! However, you would be hard-pushed to find a cleaner-running D11 anywhere. Surely it is just a matter of time before we see the use of diesel particulate filters on more

construction machines in Europe, especially those used on construction sites in towns and cities.

The next round of European Union legislation on engine emissions does address particulates, but it will reportedly only apply to engines fitted to new earthmoving machines. The Swiss emissions regulations are much stricter because they include consideration for the pollution caused by older machines as well as newer models.

While the D11T was the first Jura Cement machine to get the new system, their trucks have since received similar treatment. However, their older vehicles are slowly being replaced. While the recession has forced many European quarries to review new machine purchases, the Swiss quarry is midway through a significant three-year spending spree to renew its entire

8.11. A brand-new Cat 990H loads limestone.

8.12. Four new tyres for a 990 cost nearly €30,000 so chains are used to help boost the life expectancy of tyres from between 7,000 and 8,000 hours to as much as 12,000 or 14,000 hours.

8.13. This 100 tonne 992K is in the workshop for a service. It is the newest of three Caterpillar wheeled loaders at Jura Cement.

machine fleet. The future at Wildegg is not only bright but also yellow; this is because, with the exception of one machine, the Cat brand is the favoured option for its nine other main machines.

The favouritism began in 2009, with the purchase of one D11T, one 992K and one 990H wheeled loader and the first 775F truck. Jura Cement were so impressed that they have not only planned to part-exchange their 992C wheeled loader for another new 990H, but also to replace both an ageing Cat 771C and a dilapidated 771D.

The 771C will be the first to go. After operating for Jura for approximately 14,000 hours, the Cat truck will be part-exchanged for a new 775F some time in 2011. The 771D, which has notched up around 12,000 hours of operation, will likely follow some time in 2012. Although still at the discussion stage, mine management are considering adding a new 930 wheeled loader and a D8T dozer in 2012 in order to complete the four-year investment plan. The D11 will probably remain the only one in Switzerland because its size makes it too big for other Swiss quarries. With all the advances that are being made in environmental legislation, by the time this machine needs replacing the next model will probably feature diesel particulate filter technology as standard.

8.14. After 23,000 hours, the older Cat 992C now stands idle. The plan is to sell it next year and replace it with a brand-new Cat 990H. Next to it is the very first 992A, a model from 1971.

CHAPTER 9

Scotland

Coal has always played a vital if not dominating role in Scotland's mining history. The three main seams, from which the vast majority of underground supplies have been extracted, are located in Ayrshire, Lanarkshire, Fife and Kinross. At its peak in 1913, the 450 deep mines in Scotland produced a staggering 42 million tonnes of coal – the equivalent of 10 tonnes for every man, women and child in the country.

Affectionately referred to as Scotland's black diamonds, the country's prosperity was built off the back of the black gems. At one time, the industry provided jobs for 148,000 people – or ten per cent of the country's population – until things started to go horribly wrong. When other forms of energy – such as nuclear power, natural gas and North Sea oil – became increasingly popular, underground coal mines were badly affected.

One of the best places to learn more about Scotland's coal mining history is the Scottish Mining Museum at Newtongrange. Located just a few miles south of the Edinburgh by-pass, the museum's home is the Lady Victoria Colliery, which was one of the finest examples of a working Victorian colliery in Europe, until it was closed in 1981.

Sunk on the eastern side of the great Midlothian Coal Basin, the 550 m main shaft was once the deepest of its kind in Europe. Sadly, Lady Victoria was another casualty of the slump in demand for Scottish coal and the shaft has long been backfilled. Fortunately, the site

9.1. The Scottish Mining Museum, housed in the former Lady Victoria Colliery, provides a great insight into the country's coal mining past. It regularly receives 40,000 to 50,000 visitors a year.

9.2. The 2,000 hp winding engine was powered by steam from eight Lancashire boilers.
For almost ninety years men and coal were hauled up and down the pit at speeds of 50 km/hr.

9.3. Much of the machinery used in former times has been salvaged and is on public display.

was saved from demolition and preserved for future generations when the museum opened in 1984.

One of the highlights for many of the 40,000 to 50,000 visitors that tour the facility each year is the 2,000 hp winding engine. The twin cages hauled men and coal up and down the pit at speeds of 50 km/hr. The entire journey took exactly one minute and twenty seconds. Powered by steam from eight Lancashire boilers, even today this machine, the largest steam engine in Scotland, is still in working order, and its boilers are next on the restoration list.

Along with the winding engine, a tour of the pithead and the old power house is also a must. The accompanying story of coal exhibition is a fascinating one, providing insight into the Scottish coal mining sector and the appalling conditions suffered by many early underground miners. On display at the tour site is a 40 ft high-coal shearer, underground carriages, cages and locomotives. These machines, while massive for their time, are no match for the modern surface miners today.

Muir Dean

At one time, Scotland was the playground for a machine that was once the largest fully hydraulic excavator in the world – a 540 tonne Mannesmann Demag H 485. The prototype of this machine started life at Roughcastle coal mine in Falkirk in 1986 where it worked for Coal Contractors while coupled to a phenomenal 23 cu m bucket. The Demag H 485 was later acquired by Halls Construction and eventually exported to Canada.

While this massive excavator has long been relegated to the history books, the merger between Demag and Komatsu meant that the model specifications of the old Demag H 485 evolved into the H 655S. Eventually that old excavator design became the basis for the PC8000:

the largest hydraulic excavator currently produced by Komatsu. Made in a factory in Germany, this phenomenal 750 tonne excavator is too large for European mines, though many present-day surface mines still rely on the power of seriously large hydraulic excavators to rip up former deep mines and mop up the scraps.

One of the best places to witness the power of massive excavators is Muir Dean, a new coal-mining operation not far from the Firth of Forth bridge, north of Edinburgh and close to the village of Crossgates. Coal has been excavated at this site for centuries. Originally, the excavation process was operated on a small scale by local monks but when deep mining took off in the nineteenth century there were nearly fifty underground mines in the area. Today, the remains of the pithead frame and winding gear belonging to one of the closed mines is on public view at the nearby Lochore Meadows Country Park, which was built on land that was once covered in coal-mine waste.

Considering that these early miners were working four main coal seams – Glassee, Myneer, Fivefoot and Dunfermline Splint, all of which were lying within a 40 m band from the surface – they did a good job. Unfortunately, all that remains of these workings in many cases are the coal pillars that held up the roof.

9.4. Scotland used to be the playground for the biggest hydraulic excavator in Britain – a 540 tonne Demag H 485. In 1986, the prototype of this machine started life working for Coal Contractors at Roughcastle coal mine in Falkirk.

9.5. Nearly a hundred years ago, underground miners crawled between the coal pillars that were left to hold up the roof. Today, ATH Resources is after the remaining coal left at Muir Dean.

The site of Muir Dean is operated by ATH Resources, a company that was founded more than a decade ago by Tom Allchurch, Mike Tod and John Hodgson. ATH currently operates five surface coal mines in Scotland: Skares Road, Laigh Glenmuir and Grievehill in East Ayrshire; Glenmuckloch in Dumfries and Galloway and Muir Dean in Fife.

When ATH Resources put in their bid to open Muir Dean, there were some serious issues to address. Working and blasting times are requirements all surface mines have to obey strictly and noise-reduction systems for trucks and excavators are also essential – but the big problem was water. The site had become waterlogged and a permanent solution to the problem of the Fordell Day Level, an old drainage tunnel originally constructed to drain the underground collieries, needed to be devised. This dated back to the eighteenth century.

Originating north of Crossgates, the drainage tunnel used to surface within the grounds of the Fordell Castle, where it discharged into the Fordell Burn which would ultimately feed into the Keithing Burn. In the 1970s, when the last of the deep mines closed, water pumping also ceased, flooding many deep mines and thereby releasing iron-contaminated water into the old drainage tunnel.

The contaminated drainage tunnel traversed the middle of the proposed site, so ATH had to think of a way to remove it safely along with its poisonous water. Living up to the ATH logo of the aardvark – an African mammal well equipped for digging – the company elected to dig down to intercept the tunnel. Engineers then pumped the polluted water to the surface and into a series of lagoons. The majority of the iron settled in the sediment, while the surrounding reed beds naturally removed the last traces of iron from the polluted water. With the water issues resolved, ATH Resources was able to pursue the serious task of extracting 10,000 tonnes of coal a week, large amounts of which supply Longannet power station.

In addition to four Cat D9 bulldozers, the impressive machine fleet at Muir Dean includes one of only three rare Cat 16M graders in Scotland. Also of interest is a fourteen-year-old, German-made O&K RH120C, which was produced toward the end of 1996. The machine is interesting not only because it has worked for over 40,000 hours, but also because it features all-electronic controls and acted as a half-way stage to the newer RH120E model.

The O&K RH120C was a benchmark in excavator

9.6. Flood water from former deep mining operations still flows through an old drainage tunnel, which traverses the middle of Muir Dean.

production and unmatched when it came to shifting overburden in the British coal mining industry. The O&K factory in Germany has always viewed the tough Scottish conditions as a fantastic training ground for this excavator. Even after 40,000 hours of punishing duty it is still available ninety-six per cent of the time and will be a tough act for any new excavator to follow. Plant manager Ian Dryburgh reckons the RH120C is nothing short of amazing. 'It is a fantastic and virtually indestructible excavator. All we have to do is throw a new engine at it every once and a while and it just keeps on going.'

The secret to the longevity of this excavator is largely

9.7. After 40,000 hours of active duty, this RH120C is still going strong and the Scottish mine is confident that it will easily survive another 40,000 hours or so.

due to its solid build, but working with only the most experienced fitters and welders is also a big help. The RH120C has a daily routine whereby it stops for thirty minutes at 11am and again at 3pm so that mechanics have time to check it over. 'This is the only way to reach the high availability figure. I am convinced that with a bit of tender loving care this excavator will easily survive another fourteen years and provide another 40,000 hours of service,' says Ian.

Another way in which ATH keeps the RH120C in top condition is by utilising weekend hours and weekday evening hours (between 7pm and 7am) for maintenance. There are very few jobs or repairs that take longer than a week but when major welding jobs are necessary they are tackled during one of three annual shutdowns: two weeks in July, ten days at Christmas and a week at Easter.

The policy of ATH is to take good care of the trucks and excavators through preventative maintenance. Every ten days each of their eleven Cat 777s are given a thorough clean to check for cracks using a pressure washer designed in-house. With mechanics taking a full day to clean just one truck, the attention to detail clearly pays dividends; the trucks at Muir

Dean are in better-than-average condition. When time permits, the excavators get similar treatment.

Ian is oozing with O&K excavator experience. While working for twelve years as a mechanic and service engineer in South Africa he helped to assemble five RH200s. In 1998 he joined Scottish Coal as plant manager, taking control of thirty 250 tonne excavators. Twenty-one of these machines were O&K RH120s, with the C series model accounting for seventeen of those. Ian also had under his control a number of Demag excavators from which he gained an immense amount of knowledge. In addition to two H 185 models, two H 185S models and two H 255S models, Scottish Coal also used to run a very rare Mannesmann Demag H 285S, which tipped the scales at well over 300 tonnes.

The excavator was acquired by ATH after the purchase of Miller Mining and it came with a hefty 19 cu m bucket. Sadly, the slew ring failed at about the same time that it needed a new engine. Since it would have cost copious amounts of money to fix both problems, the excavator was scrapped. The spirit, and the basic design, of this excavator lives on in the Komatsu PC4000, which carries an even

9.8. ATH Resources has five Komatsu PC3000-6s. The first pair was commissioned at Muir Dean in the autumn of 2008, while the other two work at a site in Fife.

bigger 21 cu m bucket. Two electric front-shovel versions of this excavator work at a copper mine in Serbia.

Plant manager Ian also ended up with a massive Bucyrus-Erie 1260 walking dragline under his control. Bought in 1981, the ancient dragline was parked in 2001, just twenty years later. Despite various attempts to sell it, four years after that the Bucyrus-Erie 1260 was cut up by EP Industries. The parts were moved to India.

Appointed plant manager at Muir Dean when the site opened in 2008, one thing Ian has learned over the years is the crucial role that attention to detail plays in looking after an excavator. For example, Ian has the machines undergo regular oil sampling. He is convinced that this provides an early warning of the first sign of a problem. 'We also keep a careful watch for cracks on the boom, stick and bucket by regularly power-washing the machines. Expensive downtime occurs if you let small problems go too far.'

Supporting the RH120C are two Komatsu PC3000-6s. ATH bought five of these in total. The first two Dash 6 excavators were commissioned at Muir Dean in August and September of 2008, while the other three earn their keep alongside three other RH120s at Glenmuckloch, in Dumfries and Galloway.

The two newcomers are capable of shifting in excess of 700 cu m an hour in tough conditions and more than 850 cu m an hour in a softer dig. Notching up approximately fifty-nine hours a week, these Dash 6 excavators should last a long time.

Service and back-up are crucial with all excavators and Ian admits to being pleasantly surprised with the speed at which KMG Warrington – the Komatsu importer – has solved all issues so far. At one point, one of the PC3000s suffered from a major grease valve failure, so the component was quickly replaced on both excavators. The computer screen onboard one of the excavators presented the next challenge, but that too was easily changed by KMG Warrington. Finally, a stick cylinder on the front shovel excavator was quickly replaced after showing dangerous signs of wear.

The Dash 6 excavators are now well over the initial 5,500 hour warranty, which means that when they

9.9. The two Komatsu PC3000s did not get off to the best start; a major grease valve failure resulted in quick replacements needed for both excavators, and one of the stick cylinders on the front shovel excavator needed replacing after showing signs of wear.

encounter problems they will not be so easily fixed. It will be interesting to see whether the excavators are able to endure quite as many hours of operation as the RH120Cs they work alongside.

The PC3000 was born out of the Demag H 185, another former British mining icon, which evolved into the H 185S and then the H 255S before finally becoming the PC3000-1. Strangely, very few 185 and 255 models survive today, while numerous RH120Cs are still going strong. Ian's veteran opinion on this matter is that this is down to cost.

As an example, he estimates the cost of a new V12 engine for a Demag H 185 at around £95,000; even reconditioned engines will tend to top £53,000. By comparison, a 6-cylinder power pack for an RH120C costs just £34,000, while a reconditioned engine can be sourced for just under £20,000. The O&K excavator may need two of them but the finances still favour the RH120C rather than the expensive Demag H 185 model.

As for hydraulic pumps, a pump for a Demag H 185 costs around £15,000, while the same part for an RH120C costs less than £10,000. Another big issue with the H 185 is the cost of replacement hydraulic cylinders. Ian estimates that he can source a stick cylinder for an RH120C for under £10,000, which is half the price of a new stick for an H 185.

Some time soon the RH120C will need a major refurbishment to renew the engines, and fit reconditioned power take-off, slew ring and swing and travel motors. New track chains, hydraulic pumps, cooling fans and engine radiators will also be needed, as will a complete re-wiring of all electrical systems in contact with hot components. The refurbishment will also include copious amounts of welding and the replacement of all main boom pins and bushes – a very expensive task.

While the total bill could easily run up to £800,000, their refurbished excavator will run like new. Ian is convinced it will definitely be worth the money and, either way, refurbishment is much cheaper than spending nearly £3 million on a new excavator. So far, he has supervised two of these massive overhauls and a he finds that a crucial new addition to the revamped machines is a new seat in the cab. 'This is really important because operators think they are sitting on a new machine.'

Although the RH120C is a virtually unstoppable digger, there are two things that could result in major

downtime – the slew ring and a complete track overhaul. He reckons eighty per cent of the machines that are parked today are probably that way because one of these two component parts went wrong.

The slew ring on the RH120C at Muir Dean was reconditioned for the bargain price of £20,000. Buying a new one would have cost around £90,000, and that is the cost without the addition of the labour and machinery needed to fit it. Although the track chains wear, the biggest cost is to replace the actual pads themselves. On the RH120C there are forty-five track pads each side, which cost nearly £2,000 each. Thus, the total cost to replace the lot is a hefty £90,000 per side. 'When you have to replace track pads, the overhaul becomes really expensive.'

This is one of the reasons why many mines try to source an RH120C for breaking. Even though they often command an average price of £120,000, customers lucky enough to be able to reuse the tracks can save a fortune.

While we will probably never ever see another brand-new RH120C, some mines are concerned about the level of electronic wizardry that has crept into the newer E versions of this excavator. Ian thinks that more than half of the breakdowns to the RH120E are of an electronic nature.

Present-day Bucyrus has to pull out all the stops because no matter what the traditionalists say there are still a number of alternatives to the classic master of overburden removal. Komatsu has successfully re-entered the British mining scene with its Demag-inspired PC3000, while Liebherr is attempting to oust the RH models from Scotland. However, these new contenders will have to prove the test of time, something that the RH120 has managed with style and grace.

Liebherr R 9350

Located within sight of the M74 in South Lanarkshire, Broken Cross is easily the largest coal mine in Scotland. Coal excavated from the site makes up a quarter of the 4 million tonnes of coal excavated by the Scottish Resources Group (SRG) through its subsidiary Scottish Coal.

Broken Cross was once the site of the Douglas Water Colliery. Estimates suggest that there is 13 million tonnes of coal buried below the surface; therefore, the removal operation should keep Scottish Coal busy for another eight years. During this time the big machines will move a staggering 250 million cu m of overburden, ultimately

9.10. Broken Cross is the biggest coal mine in Scotland, producing around a million tonnes of coal a year.

causing the current depth of the void to drop to 200 m.

Like many other Scottish surface mines, the thirteen coal seams at Broken Cross are not particularly deep. The smallest seam has a thickness of just 10 cm, while others are as thick as 60 cm. Fortunately, Broken Cross is positioned over the remains of a 9 ft (2.75 m) seam from a former underground mine. The early miners did

not manage to exhaust this seam, so Scottish Coal gained another area from which to extract coal.

Most of the coal seams are separated by between 6 and 7 m of overburden – though sometimes it can be as deep as 20 m. On average, excavators must shift 25 cu m of overburden to uncover just 1 tonne of coal. To reach the 22,000 tonnes of coal needed every week operators must

9.11. Blasts are carried out at least once a day and on average the mine gets through 60 tonnes of explosives a week.

9.12. Broken Cross is the home of the only five Liebherr R 9350s in the country. Each one of these powerful excavators weighs 320 tonnes.

move a phenomenal half a million cu m of overburden.

Historically, most of the surface mines owned by Scottish Coal worked with two O&K RH120C excavators. Castlebridge Plant – SRG's in-house machinery company – has twenty-two of these older excavators, many of which have notched up an astonishingly high number of hours: some in excess of 75,000. However, these machines are slowly coming to the end of their lives which is a problem because parts are scarce and difficult to source. Now that Bucyrus is at the helm – and the prospect of a takeover by Caterpillar is a possibility – companies that own these machines hope that things will change and that improvements will be made. (See Postscript.)

Demag was another former popular excavator with Castlebridge Plant, but their sole surviving H 185 will probably be retired soon. New excavators are slowly being added to renew the loading fleet but rather than place all their eggs in one basket the management at Broken Cross currently follow a parallel path when it comes to purchasing new excavators. After an exhausting selection procedure, Scottish Coal took the plunge and invested heavily in five of the biggest and

newest Liebherr excavators in the country: a move that is considered possibly the most comprehensive renewal programme of a machinery fleet that the British mining industry has ever seen.

In the past, Liebherr sold some fairly large excavators in Britain. The biggest active survivor is a 230 tonne R 994, which features later in this chapter. However, even this former giant earthmover is no match for the five 320 tonne R 9350s – the largest Liebherr excavators in the country – that were purchased by Scottish Coal during the massive renewal. Four were commissioned in early 2009, while the fifth machine arrived in the middle of 2010. Worth an estimated £10 million, the initial order for the first four by Castlebridge Plant remains one of the largest single excavator orders ever placed with Liebherr by a company in Britain.

The new French factory at Colmar is home to the production of the R 9350 but two of the excavators were made at the Brazilian Liebherr factory in Guaratingueta. When they reached British shores, sixteen low loaders were needed to transport the components of just a single machine to the mine.

Things started badly for Scottish Coal's new Liebherrs

9.13. The R 9350 is the successor to the old R 994B. The five excavators at Broken Cross are divided between three backhoe and two front shovel versions; all are fitted with 1,500 hp engines.

when one of the exhaust valves on one of the massive 12-cylinder engines snapped just a few minutes after turning the key. Though the exhaust valve succeeded in dropping into the cylinder and wrecking the piston, Castlebridge was impressed by Liebherr's prompt response to the problem. If the company had not managed to provide a quick solution, then it could have spelt disaster for the multi-million pound contract. Luckily, a brand-new replacement engine arrived at the mine the next morning and the giant excavator was back at work by the end of the afternoon!

Powered by a single 45 litre Cummins engine, which is capable of running at a maximum of 1,500 hp, the Liebherr excavator oozes power. Also there is no straining or nodding when the bucket is drawn into the rock. All five Liebherr excavators – three backhoes and two front shovels – use the same width and 17 cu m capacity bucket to ensure minimum overspill when loading the trucks, which have a capacity of 91 tonnes. Effortlessly digging up 30 tonnes in a single go, the Liebherr needs just three passes and the trucks are on their way back to the tip.

From Monday to Friday, all five excavators work two 12-hour shifts per day. Two also work for six hours on Saturdays. Scottish Coal would like to have all the excavators working on weekends but they struggle to find drivers. The remote location coupled with the long shifts – all operators work sixty hours a week – adds to the difficulty of finding employees. Fortunately, there are still some young people keen to sit behind the controls of one of the biggest excavators in Britain. Derek Mckee is the youngest at just twenty-eight years old.

Safety is of paramount importance and it comes before production at Broken Cross, with management going to great lengths to reduce any chance of an injury. Haul roads are dangerous places so where possible traffic is restricted to the big trucks. Safety is further improved by a dedicated service road running from the main office and workshop buildings to the supervisor's lookout. The service road is almost 1.5 miles long and it provides a quick, safe link between the two without service vehicles needing to negotiate the haul roads.

Xenon lighting and flashing beacons improves safety around the large excavators. These flashing warnings are engaged automatically when the machine travels, making

9.14. The excavator's 17 cu m bucket has no trouble filling one of the 136 tonne capacity trucks in five comfortable passes.

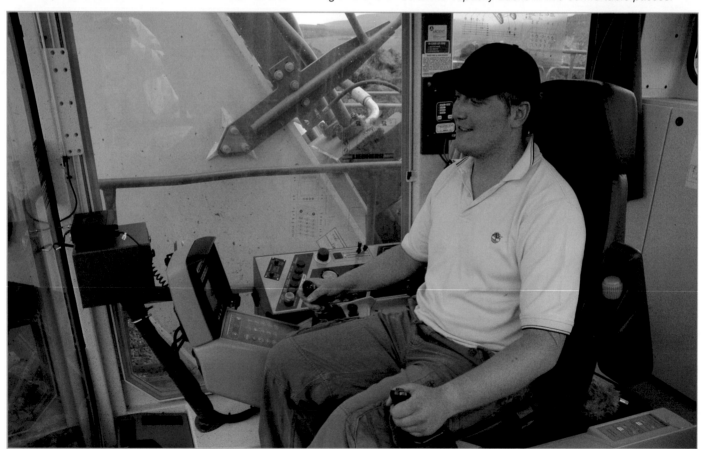

9.15. The remote location and long hours make it difficult to recruit operators. At twenty-eight, Derek Mckee is the youngest.

9.16. Each excavator stops for two 30-minute breaks per day. During these breaks, the 6,000 litre fuel tank is topped up and the machine is checked over.

them very useful at night. To further improve safety, the big excavators are serviced twice a day during thirty-minute breaks.

Travelling in convoy, the service teams are not allowed onto the haul roads until the last of the big trucks are safely out. During the short service breaks, the whole mine seems to race against the clock. Topping up the 6,000 litre diesel tanks – which drain at a rate of 190 to 195 litres per hour – and any other service-related jobs must be completed within a thirty-minute window. The haul trucks are not allowed to start their engines until all service vehicles are out of the mine and the all-clear is given.

Caterpillar 785D

Shifting more than half a million cu m of overburden a week is a truly mammoth task and one which requires a massive truck fleet. Until recently, the Broken Cross fleet included a batch of Cat 777s and twenty-six Terex TR100s – each with a capacity of 91 tonnes – that were made just 30 miles away in Glasgow. These machines were the largest and newest Terex fleet in Britain.

However, these nifty trucks are no match for a batch of even larger and newer Caterpillar trucks. The purchase of new Caterpillar 785s, the first in the country for the past fifteen years forms part of a massive £45 million

9.17. The mine used to run twenty-six 91 tonne capacity Terex TR100 trucks.

*9.18. Scotland's largest coal mine recently bought ten brand-new Cat 785Ds with capacities of 136 tonnes.
Each of these trucks cost nearly £1.4 million.*

modernisation programme that Broken Cross is undergoing. So far the Finning factory in Glasgow has delivered 125 new machines to the site in order to replace the ageing fleet.

The new trucks are covered by a seven-year or 36,000-hour repair and maintenance contract, which included a complete overhaul of each truck at 20,000 hours. The initial build of the Cat 785s was carried out at two locations: the chassis cab was assembled at Finning's Glasgow branch from the parts carried by four articulated truck loads that were dispatched from various production sites in the US via the Liverpool Docks. Meanwhile, the skips were built at the Finning headquarters in Cannock before being driven north to Broken Cross for final assembly. The reunited pieces of the massive trucks resulted in a machine measuring 7 m wide and over 11 m long.

Two 136 tonne capacity haulers carry the same volume of material as three of the smaller trucks; a fact that not only allows Broken Cross to reduce the number of vehicles pounding the haul roads, but also helps them to reduce the number of employees by five drivers per shift.

With fewer trucks to maintain the mine is saving an absolute fortune on diesel. For example, a 91 tonne truck burns 70 to 75 litres of diesel an hour. Though the 785 is powered by an immense 12-cylinder 1,450 hp Cat engine, fuel consumption is only marginally higher at 105 to 110 litres per hour, which helps to reduce the cost per tonne.

The arrival of the largest new Caterpillar trucks in Britain finally provided the massive Liebherr excavators with a challenge; subsequently, production figures increased immediately from an average of 850 to nearly 1,000 cu m per hour solely due to the five- rather than three-pass fill and reduced truck positioning times. A 136 tonne truck is also a significantly bigger target for their excavator buckets, helping to make filling the skips a much more relaxing task.

9.19. Fitted with an 8 m wide blade, this 550 hp Cat 834H was the first in Britain. It weighs 50 tonnes.

9.20. Not far from Broken Cross another coal mine – called Broken Cross North Extension – is taking shape. This new void keeps this old O&K RH40 fully occupied.

9.21. This twenty-year-old Cat 777, the mine's main dust eater, makes quick work of damping down the haul roads.

Safety First

Big trucks are notorious for driver visibility – or rather lack of it. When the first Terex arrived on site, the mine looked at ways to improve visibility for the operator. Working with Spillard Safety Systems, Scottish Coal was keen to provide their drivers with an all-round view of the truck – one metre above the ground. Visibility from the seat was good, but drivers complained that the view to the steps at the front of the radiator was obscured. It turns out that this problem was easily solved by fitting a convex mirror to the front of the bonnet. Then the small mirrors on each side of the truck were swapped for larger ones allowing drivers to see a much greater area.

With the mirrors sorted the attention turned to the cameras. The TR100 was already fitted with a standard rear-view camera so they added a second one to the offside handrail, with images from both cameras

displayed on separate CCTVs. The inclusion of new mirrors and an extra camera eliminates all blind spots, providing drivers with great all-round visibility, but as an extra precaution reflective tape and warning text is added to the sides of the skips. Once management was satisfied that everything possible had been done to improve safety, the measures were gradually added to the rest of their Terex truck fleet. The mine declares that if the costly investment prevents just one serious accident then it will have paid for itself.

Their big new Cat trucks did not escape the safety tweaks and, after a week of experiments, Broken Cross got Finning to relocate the automatic fire extinguisher to eliminate a blind spot. The positioning of the lights was also altered: not only did they dazzle oncoming service

9.22. Scottish Coal not only has the only R 9350s in Britain but also the only R 9250s.
Their four 250 tonne excavators are each fitted with 15 cu m buckets.

vehicle drivers but they were always dirty, requiring constant cleaning. Both issues were solved by moving the lights up from the bumper to a position on the radiator grille.

Operators were less than impressed by Caterpillar's standard format for having three camera images appear on a single CCTV, so they did away with the camera at the top of the access steps and replaced this with a mirror. In the cab, there are now two screens showing continuous images of the rear-facing and offside cameras.

The side mirrors were next to come under scrutiny and the large traditional one-piece mirrors on both sides were swapped for the modern two-piece design from the 777F. The modern mirrors provided the operator with a better degree of forward and side visibility. The replacements are also heated, which is a big bonus in the harsh winter climate.

Finally, mounted on the front guard rail, three close-proximity mirrors not only provide a view to the operator's side, as well as the centre and offside, but also to the front of the truck. This allows the operator to see even when someone is standing right up against the bumper.

Liebherr R 9250

Scottish Coal must have been pleased with the performance and service of the five R 9350 excavators at Broken Cross because they recently bought two more slightly smaller R 9250 Liebherr excavators. These excavators were originally launched to succeed the R 994. The first 250 tonne R 9250 backhoe excavator arrived at Powharnal in Ayrshire in November 2009 to replace an older O&K RH120C. The R 9250 worked

9.23. It takes two weeks to fully assemble a brand-new R 9250. Once the air-conditioning and automatic fire extinguisher systems have been added this one will be ready to make a start.

9.24. Two of Scottish Coal's R 9250s work at House of Water, a site that has yielded 6.5 million tonnes of coal since 1996.

9.25. The first R 9250 started work at House of Water in April 2010. It has since been joined by another and both are fitted with a 9 m boom and 4 m stick.

alongside two O&K RH120Cs but it did not stay long at this site because coal production wound down in the autumn of 2010 and the Liebherr excavator was moved to a neighbouring site called Ponesk.

Site manager Alan Hiddleston admits he was more than a little apprehensive when the first R 9250 arrived. 'We are used to red and white machines, a tradition that is difficult to break.'

While many of his operators were not overly keen to sit behind the controls of the new Liebherr excavators, these early reservations soon turned to respect and now the operators are all big fans of the French-made machines. Hardly surprising considering operator comfort has massively improved from the classic machines they previously controlled.

Alan still feels that the digging performance of the R 9250 does not quite match that of the RH120, but he admits it is not as far off the mark as he previously envisaged. However, he is clearly still very much an O&K fan and old habits die hard. 'I view the Liebherr excavator in much the same way as one might view a family saloon car: safe and a little boring. The RH120E, on the other hand, is a sports car.'

Despite early reservations, Castlebridge Plant continues on a parallel path to add both new RH120s and more new Liebherr excavators. In fact, Scottish Coal recently purchased another two R 9250s for the House of Water surface mine.

The East Ayrshire area, which encompasses House of Water, is the most important coal-producing area in Scotland today. Good for around 3.5 million tonnes of coal a year, four of the six main surface mines in the area are currently operated by Scottish Coal. Although

9.26. Power comes from a 45 litre V12 Cummins engine that generates 1,287 hp. It has a fuel tank capacity of 5,440 litres.

the decrease in demand for coal resulted in the closure of House of Water for two years between 2005 and 2007, the site is now moving full speed ahead. Since it first opened in 1996, House of Water has relinquished over 6.5 million tonnes of coal. The deepest of thirty-four coal seams at House of Water lies at 134 m below the surface and the thickest seam, which outcrops at the surface, is 2 m.

The mine prides itself in being the only one in the country to have a river running right through the middle of it. This body of water is called the River Nith and its salmon-spawning waters have been diverted twice to get at the coal buried below. The first re-route

9.27. Scottish Coal uses nearly thirty dozers and four 580 hp Cat D10Ts.

occurred in 2000 and the second in 2004; these were the biggest salmon-spawning river projects in Europe. The last move, six years ago, saw the river put back close to its original course.

Another fact that House of Water mine management takes pride in is that ninety per cent of their employees live within a 20-mile radius. Whenever possible they try to recruit local people, a sort of token gesture to show they are trying to give back to the community. 'Recruiting staff is not easy at the best of times, especially when it is restricted to the local community,' says site manager Gary Ward. To assist with recruiting the mine runs an internal training programme that is open to people with no previous machine experience. The programme is a big commitment for both trainee operators and mine production since it takes three weeks before an operator can be let loose with one of the trucks.

In terms of excavation, House of Water used to rely on the services of three RH120Cs and a twelve-year-old RH120 – one of the very first E series versions of this excavator. This machine was plagued with no end of problems from hydraulic pumps that had to be replaced to slew rings being changed three times during the first

16,000 hours of operation. Once the Liebherr excavators arrived on site, mine management was happy to move the temperamental RH120E. The RH120C models were taken out of service and refurbished and then moved elsewhere.

Transporting the RH120C unveils another big feature of the model: each excavator can be easily split, over a period of about two days, into six loads. One vehicle is needed for the boom and bucket; a second for the stick and ballast; and a third for the main body. The fourth and fifth vehicles can carry the track frames, while the sixth takes the undercarriage and cab. Once the excavator has reached its destination, the rebuilding process only takes two days.

Each excavator is fitted with a newly designed 15 cu m bucket and is covered by a 15,000-hour repair and maintenance contract, after which they will assess the condition. In the meantime, the machines are expected to clock up 110 hours of active duty a week. The new bucket design was essential for working in the tough conditions on the House of Water site. The coal is concealed by a 33 m layer of some of the country's toughest whinstone, highly abrasive and capable of destroying bucket teeth after just a few days of intense work. Excavators are

9.28. The O&K RH120C is a mining legend in Britain and while Castlebridge Plant continues to renew its machine park, it still runs the biggest fleet of 120Cs in the country.

expected to strip this overburden at a rate of 230,000 cu m every week in order to uncover 10,000 to 15,000 tonnes of coal.

The deep and narrow bucket design, which is also nearly 1.5 tonnes heavier than the standard bucket, provides the Liebherrs with similar breakout forces to the RH120E. The newer excavators are capable of moving an average of 650 cu m of material an hour, which is approximately the same as the RH120E. In order to extract this high volume, the Liebherrs need to use just under 170 litres of diesel an hour, again similar to the usage of the Terex excavator.

After a bit of initial hesitation that is typical of major change, the House of Water operators seem to have warmed to the French-made excavators. The feature that gradually persuaded the operators is obviously the comfortable operating environment. All minor teething problems aside, the Liebherr excavators appear to be living up to their potential and have established themselves as worthy competitors to the RH120.

Bucyrus RH120E

The O&K RH120 has been a popular addition to Britain's coal mines, with many still going strong after twenty years of service and more than 75,000 hours of operation. The longevity of design, a key feature in the excavator's success, was left largely unchanged after Terex Mining bought the O&K division in 1998. In the decade that followed, both old and new RH120s alike have built up an enviable reputation as solid and steady workhorses.

The business side of building these excavators, however, can be described as anything but steady. The tumult all started in early 2010 when Bucyrus beat Caterpillar to the post in the race to acquire Terex Mining. Senior management at Caterpillar reacted strongly to this development and quickly revealed plans of multi-million dollar investment to develop a full range of American-made mining shovels ranging in weight from 125 to 800 tonnes.

As if that was not enough, toward the end of 2010 the

9.29. *The first pair of RH120Es to wear the Bucyrus livery in Britain were actually made in 2005 but the unusual livery may be short-lived if Cat's bid for these excavators is successful.*

tale took another twist when Cat rocked the global mining community to its foundations with a whopping $8.6 billion bid to buy the complete Bucyrus business. This acquisition could soon give Caterpillar mining excavators such as the RH400, which at 980 tonnes is the largest excavator in the world, as well as access to monster electric rope shovels, draglines, drills and trucks. (See Postscript.)

The first seriously large Caterpillar mining excavators since the demise of the ill-fated Cat 5000 series in 2004 could well be a reality soon. Ahead of regulatory approval, which is expected toward the end of 2011, the new Bucyrus livery – white, dark grey and purple – that adorned the first batch of Bucyrus RH120Es could be a thing of the past.

The very first pair of Bucyrus RH120Es to work in Britain were originally the property of a Venezuelan mine, which purchased them back in 2005. However, shortly after arriving in the country the South American mine went bust and the two brand-new excavators spent the next five years sitting idle at the docks. Eventually, the excavators made it back to the factory in Germany before they were snapped up by Castlebridge Plant – long-term users and fans of O&K. At the same time when Castlebridge Plant purchased these excavators, they did another deal for two more as part of a massive investment to modernise its extensive fleet of mining machinery.

The first two worked at Ponesk, a relatively new coal mine that was previously known as Spierslack. They were contracted to Banks Mining by Scottish Coal and when the contract expired, Scottish Coal took full control of the site. They put the two Bucyrus RH120Es to work helping to complete the last section. The big excavators are expected to remove over 23 million cu m of overburden within the next five years.

With 120,000 cu m of overburden a week currently being extracted, coal production is still in the early stages with just a few thousand tonnes leaving the site each week. Scottish Coal needs another couple of excavators to take out the remaining million tonnes of coal. All of the coal mined from Ponesk goes to Killoch, which was one of the largest underground mines in the days of British Coal. Today, it is the name of a rail disposal head.

The two RH120E excavators are identical with the exception of their buckets: one carries a 15 cu m bucket, while the other is connected to a 4 m wide, 17 cu m version for use in light soils. Loading Cat 777 trucks, mine management expects each excavator to shift an average of 700 cu m per hour.

Their operators love the newer versions of this excavator, which is something because the RH120C version was a tough act to follow. Often seen as the best machine the British coal mining industry has ever produced, the RH120C will be difficult to follow in terms of endurance and availability. This popularity had much to do with the simple mechanical design of the machine: very few Scottish mechanics have not worked on an RH120. The RH120E version, on the other hand, hides more complex electronics. Many mine managers admit they would love to be able to get their hands on a brand new RH120C today.

Castlebridge Plant is keen to see whether these newer

Bucyrus RH120Es will provide the same twenty years of reliable service and between 70,000 and 80,000 hours of operation that the RH120Cs managed. In 2009, Bucyrus created a prototype of an RH120E that was fitted with a 27 litre Cat C27 engine in Egypt. This proved to be a sizeable increase on the initial 18 litre C18 engine of previous models. Just a year later, the first two excavators fitted with the bigger engine started work at a coal mine in England.

Scotland's Biggest Quarry

Dunbar Works, owned by Lafarge, is bordered by Firth of Forth to the north, the Torness power station to the south, the North Sea to the east and the A1 to the west. Not only is the site the only cement plant in Scotland but it is also the country's biggest quarry. Roughly 30 miles east of Edinburgh, and a similar distance from the English border, the location affords great views of a famous local landmark called Bass Rock.

The quarry was started by Blue Circle Cement in 1960, and though ownership has changed hands, many of their inherited prime movers have been around since the

9.30. Scotland's largest quarry was operated by Blue Circle Cement in 1960 but is now owned by Lafarge.
At full production they shift 4.5 million tonnes of material a year, 1.4 million tonnes of which is limestone.

9.31. The 1999 Demag H 135S excavator spends much of its time drop-balling, reducing 10 to 20 tonne blocks to manageable pieces in a matter of minutes.

beginning and are true dinosaurs. Dunbar Works has a 140 tonne Demag H 135S which, after 26,000 hours in some of the toughest material imaginable, continues to survive the punishing conditions.

Only five of these excavators ever came to Britain and of the five, one was shipped to Norway, three are unaccounted for – presumably scrapped – and the fifth is at Dunbar Works. The fifth excavator was the last to enter Britain and is quite possibly the only one still working today. Commissioned in 1999, the Demag H 135S is popular with operators and is still relied on to work two shifts per day, from 6am to 9pm.

Truck loading is not the only task for the excavator's 9.5 cu m clamshell bucket, it also spends time drop-balling for secondary breakage of large lumps, pulverising blocks weighing 10 to 20 tonnes in just a few minutes.

While the excavator is clearly in the last stages of its

life, it was recently renovated with a new Cummins KTA38 engine. It received a badly needed hydraulic refurbishment and all main bushing points were cut again. The whole procedure cost nearly £160,000 but management is confident that it will provide at least two more years of service. If the quarry wants to replace it, however, they may run into trouble. When Komatsu acquired the Demag H 135S, it became the PC1400, an excavator that is no longer available. Dunbar Works may have a dilemma if management wants to stick to Komatsu excavators because the PC1250 is probably too small for their operation, while the next model in the PC family, the PC2000, may be too large.

Fortunately, Dunbar Works still has plenty of time to consider the options. Not only will the Demag soon need replacing but so will their second prime mover, a Liebherr R 994 from 1996. This second machine was one of just eleven of these excavators sold to companies

9.32. This 1996 Liebherr R 994 is believed to be the last survivor of eleven brought to Britain.
Fitting it with a shorter stick allowed the excavator to increase bucket capacity from 6 to 10 cu m.

9.33. The Liebherr excavator fills Terex TR100 trucks, which are gradually being phased in to replace the twenty-year-old Cat 777Bs.

9.34. The Dunbar quarry used to be home to what is believed to be the only Bucyrus-Erie 380-W in the country.
This picture was taken in the summer of 2010 but before the year was out this machine was cut for scrap.

in Britain and it is one of three bought by Blue Circle Cement for the massive Northfleet quarry in Kent. The Liebherr R 994 has a slightly shorter stick allowing it to carry a 10 cu m bucket, bigger than the factory-fitted 6 cu m, and better for filling the 91 tonne capacity trucks.

The pump drive gearbox on the Liebherr was renewed in October 2008 and the engine was swapped in May 2009 after the excavator had reached 16,000 hours of operation. It is currently just above the 21,000 hour mark, so Dunbar Works have budgeted to replace it in a couple of years for a similar-sized backhoe excavator. However, management is quietly confident that the Liebherr excavator, which is completely mechanically sound, could remain operational for a bit longer if necessary.

Old as they are, the two prime movers are mere fledglings when compared to the other giant earthmovers on the site of Dunbar Works. Up until very recently, the skyline around the quarry was intruded

upon by the massive 50 m boom of a 400 tonne Bucyrus-Erie 380-W walking dragline. This legendary relic was the last survivor of six of these walking draglines that worked at various sites in the UK.

Coupled to a 22 cu m bucket, the sole task of this awesome metal giant was to shift sandstone. Interest was expressed at the idea of giving it a new lease of life elsewhere in Britain, but the costs of moving the great machine, as well as rebuilding it ready for action, were deemed prohibitive. In the end, Lafarge concluded that the best option was to fire up the gas gun since the walking dragline was estimated to be worth between £20,000 and £30,000 as scrap metal. Within a few weeks, the last surviving machine of its type was cut to pieces and carted away.

Dunbar Works quarry is estimated to run for another thirty years, which is the time it will take to extract the remaining 45 million tonnes of limestone. However, the

9.35. The future of this Bucyrus-Erie 195-B, a former British mining icon, hangs in the balance. It is quite possible that it will be cut up for scrap before the end of 2011.

material is by no means easy to get at because it is covered by nearly 120 million tonnes of overburden. The first 5 m layer is concealed below 26 m of glacial drift, boulder clay and shale, while the second 7 m band lies beneath 10 m of sandstone.

At the moment, the process is to truck and shovel everything but, as the depth of the top overburden layer increases, current stripping ratios of 2.4:1 are expected to rise to 3:1, which might mean that the machine fleet will have to grow as well. In an attempt to eliminate the need for any additional trucks, Dunbar Works considered refurbishing the Bucyrus-Erie 195-B to rejoin

the fleet as a prime mover and assist with stripping off overburden above the first limestone layer. Made in Lincoln in 1980, the 350 tonne rope shovel was purchased new alongside the 380-W. Quarry management did toy with the idea of bringing her back to her former glory. The plan was to use the excavator's 13.6 cu m dipper to load directly onto an unusual bridge transporter system – a sort of cross-pit conveyor.

The bridge transporter system was designed in Germany by Wesserhutte and built on site in 1981. There was once a similar system at a coal mine in north-west Spain; today, however, the large metal constructions

9.36. The Scottish quarry was hatching plans to buy a brand-new excavator to load this unusual bridge transporter system. However, the downturn in the economy means these plans are now on hold.

parked at the top of the Scottish quarry are quite possibly the only ones in the world.

Refurbished by ThyssenKrupp between 2007 and 2008, the novel overburden extraction technique has not worked since; consequently, plans to revive the rope shovel were quashed after the quarry received a hefty £1.8 million quote to get it ready for action.

The fact is that the main generators on the 195-B are still in good working order but the electrics and controls have had it. The saddle-block, a component that allows the handle to move in and out on the boom, also needs renewing, which is a big job because both

dipper and handle need dismantling. The massive quote included funding for new tracks linkages and idlers as well as new hoist, crowd and dipper trip ropes.

With any hope of a revival of their 195-B now crushed, the odds are high that it will end up as scrap. For the immediate future it looks as though Dunbar Works will continue to truck and shovel everything. Management considered buying a brand-new electrically operated 250 tonne excavator to load the bridge conveyor's crusher to the capacity of 1,200 tonnes per hour; unfortunately, the economic downturn has had a negative impact on the demand for limestone

9.37. This Cat 992C really needs replacing. Originally sourced from Caulden quarry, the plan was to relegate it to the role of support machine, but it still sees more action than management would like.

9.38. Lafarge spent a small fortune returning the 1999 D10R to nearly new condition.
As well as a track and transmission overhaul, it was given a refurbished engine and radiator.

and for the foreseeable future the plans to expand have been put on the back-burner.

The fate of another of the quarry's former prime movers – the Cat 992C wheeled loader – is also uncertain. After 32,000 hours of punishing duties, the plan was to relegate it to the role of support machine. However, its 8 cu m bucket still sees more action than a wheeled loader of its age should, but without it quarry manager Dave Hurcombe admits that Dunbar Works would have a big problem. 'The Cat 992C is a major workhorse and we really need to replace it as soon as possible,' he says.

The quarry wants a newer version of the 992 but this is another difficult call because of the cost. Even a used 992D that has worked for 20,000 hours will cost Dunbar Works the best part of £300,000 and they have already spent considerable amounts of money to return their 1999 D10R to nearly new condition. The D10 was lucky: as well as a track and transmission overhaul, it was fitted with a refurbished engine and radiator and is currently awaiting new push-arms. Dave stresses the importance of the 28,000-hour Cat D10R to the Dunbar Works operation. 'It does everything. Worst case scenario: no dozer equals no roads.'

CHAPTER 10

France

A few decades ago, many French earthmoving contractors relied heavily on draglines. Today, however, draglines are not the machine of choice: the largest remaining active machine – a Marion 7400 – works at a gravel pit near Lyon. The only other working survivor of its kind is found at Aitik copper mine in Sweden (see Chapter 13). The fact is that the future of these two Marion walking draglines and that of the last remaining trio of Lima 2400B crawler draglines in France is highly uncertain.

The draglines were superseded by large hydraulic excavators from the likes of Poclain, Liebherr, Cat and O&K. These machines were popular at many of France's largest coal mines, such as Decazeville, Alès, and Montceau-Les-Mines. Decazeville used to have a Cat 5130 – the largest Cat excavator ever brought to Europe – and a Poclain EC1000 – one of the largest Poclain excavators ever made. Alès was not only home to numerous large Liebherr excavators but also to an O&K RH120C and an even bigger O&K RH200 – a rare beast in Europe. Montceau-Les-Mines relied on large mining excavators such as the Liebherr R 991 and R 992 and the Poclain 400 and 600.

One of the best places to visit in order to get an insight into some of the popular mining machines used in France is at a former coal mining operation run by Houillères de Bassin Centre Midi. This company used to operate a number of large surface and underground coal mining operations near Carmaux, which is located not far from

10.1. Although this Marion 7400 walking dragline still works occasionally at a gravel pit, the future of this machine – the last one in France – hangs in the balance.

10.2. This Bucyrus Erie 295-BII was one of the main prime movers at the Carmaux surface coal mine. Today, it is one of a number of big earthmovers on display at the aptly named Parc des Titans.

Toulouse. Opened in 1985, one of its largest coal mines contained estimated coal reserves of 15 million tonnes and the mining operation was estimated to run for thirty-five years. However, enormous operating costs and a change in direction for the country's national energy policy led to its demise in 1997, just twelve years after it was opened.

The water-filled void that used to be the centre of the mining operation now forms part of a multi-million pound theme park, complete with water sports and an artificial ski slope, called the Cap' Découverte. Open from April to September, many of the retired prime movers left over from the previous mining operation – including a Demag H 285, a Bucyrus-Erie 295-BII, several Dresser Haulpak trucks and one of three 400 tonne bucket wheel excavators – are rusting away nearby at the aptly named Parc des Titans. In fact, just a short distance from the park a mining museum offers many metres of tunnels for visitors to get a glimpse of what it was like to work underground.

While many of France's former large mining machines have long been scrapped or moved overseas, one machine that is still relied on heavily is the motor scraper. It is estimated there could be as many as a hundred older Cat 631s (of various versions) being pushed by scores of Cat D10 and Komatsu D375A dozers. At the Couvrot limestone quarry, owned by Calcia, Caterpillar scrapers are the preferred choice; this site is home to four of the largest and newest 657 twin-engine scrapers (two 657Es and two 657Gs) in France.

Today, however, France is best known in international earthmoving circles as the birthplace of the Poclain brand – to be discussed later in this chapter – and as the present-day home of many large Liebherr mining excavators. The first massive Liebherr machine was made at a factory in Colmar, located in the Alsace region, in 1966. This machine was called the RT 1000; it weighed 21 tonnes, was powered by a 130 hp engine and was fitted with an enormous 1.2 cu m bucket!

10.3. The Couvrot limestone quarry runs four twin-engine Cat 657 scrapers – two 657Es and two 657Gs. The combined efforts of these scrapers and three large tracked dozers generate approximately 1 million tonnes of cement each year.

The next Liebherr excavator in France came in 1979 with the launch of the 165 tonne, 710 hp R 991 HD. Some of these excavator models – with their 6.3 cu m buckets – still exist in mines around France today: a true testament to their durability. Many of the design features included in this older model are apparent throughout Liebherr's range of modern excavators. Even the R 9800, which at 800 tonnes is the company's newest and largest mining excavator, still resembles the ancient R 991 HD from 1979.

Wielding a massive 42 cu m bucket and packing a powerful 4,000 hp engine, the R 9800 heavyweight is one of eight mining excavators to come from a brand-new factory. The site of this factory is located at the town's regional airport just a few kilometres from the existing plant at Colmar. By building two factories, Liebherr plans to separate its standard earthmoving and mining divisions which will allow it to focus more

carefully on what it produces for each. In fact, Liebherr is currently on target to meet forecasts of 250 excavators during 2011. There are plans already underway to boost excavator numbers further in the coming years.

The new factory builds all current Liebherr mining excavators from the 120 tonne R 984C to the new, top-end R 9800. All the models have undergone a number of refinements over the years and brand-new, lesser-known excavators are sent for field-testing in secret locations before release to customers. For example, the 757 hp 110 tonne R 9100 premiered at the 2010 Bauma show, but one of the prototypes is under evaluation at a secret location in Southern Europe and will not be released until vigorous testing is completed.

The brand-new R 9400 – an excavator weighing 355 tonnes – is Liebherr's latest massive miner to undergo field-testing and be cleared for sale. This machine is capable of carrying a maximum bucket of 22 cu m on a

10.4. Although this R 991 dates back to the late 1970s, the exterior lines of this machine can still be found in many present-day Liebherr excavators. It has an operating weight of 165 tonnes and is one of just a handful of survivors in France.

10.5. With an operating weight of 800 tonnes, the R 9800 is Liebherr's largest hydraulic excavator. Though it was built at Colmar, the chance is remote that these monster miners will ever work in Europe.

10.6. This 100 tonne R 9100 is the new 'baby' in Liebherr's mining excavator range. It was launched at the 2010 Bauma show.

backhoe option or 21 cu m on a front shovel. The first two were recently delivered to Australia. Sadly, there are not any R 9400s in Europe yet.

While the prototype of the very first 670 tonne Liebherr R 996 was evaluated at a quarry in northern France for several months, there are no land-based variants of Liebherr's second largest excavator in Europe. Neither are there any of its freshly-launched successor – the 676 tonne R 966B. The only land-based 450 tonne R 995 in Europe features in the Spanish chapter.

The former factory at Colmar also used to build the R 994 and R 994B models. The numbering on these excavators may suggest that these are similar machines, but this would be a false assumption. The R 994 weighs just 230 tonnes while the R 994B weighs 297 tonnes: quite a big difference in operating weight. Eventually the confusion caused by this proved reason enough for Liebherr to bring more clarity to the two excavators by replacing the R 994 name with the new name R 9250. Liebherr's factory in Brazil may still make the R 994B but the successor coming out of the French factory is called the R 9350.

Of these newly named excavators, the very first R 9250 delivered anywhere in the world works at a Holcim quarry in the north of the country. It is also the only one of its kind to be found in France. Liebherr, which is keen to sell more of both excavators in France, may not have to wait too long for mine managers to make the switch because the handful of older working R 994s and R 994Bs are coming to the end of their lives. The best known examples of these older models are still working high in the peaks of the French Pyrenees.

White Gold

Hydrated magnesium silicate is the prime ingredient in the soft body powder that is known as talcum powder. However, this is not the only product that can be made using this mineral compound. Global demand for the substance is over 5 million tonnes a year, meaning that there are many other important industrial uses for the softest mineral on the planet.

Fifteen thousand years ago cave dwellers used it as an ingredient in their paints; similarly, talc has been used in cosmetics for thousands of years. Still used in paints

10.7. Trimouns is the largest surface talc mine in Europe and it has the stunning beauty of the French Pyrenees as a backdrop. The site produces 400,000 tonnes of talc a year, which is equivalent to eight per cent of total global production.

and cosmetics today, its resistance to heat, electricity and acids makes it ideal for many other applications including electrical switchboards, animal feed and plastics. It is also used in olive oil processing, paper and even chewing gum.

Producing 1.4 million tonnes of talc a year, Rio Tinto Minerals (RTM) is not only the world's largest single producer but also the owner of the largest talc mine and processing facility in the world. The Trimouns mine, where RTM extract the talc, is located high above the village of Luzenac in the Ariège valley of the French Pyrenees. Each year around 400,000 tonnes of 'white gold' are extracted here; this number amounts to thirty per cent of the talc produced by RTM and eight per cent of total global production. The excavating process is slowly eating away the three peaks surrounding Luzenac.

The tens of thousands of tourists who visit the area each year come to hike, climb or cycle during the day and then enjoy a soak in one of the numerous thermal springs: Ax-les-Thermes being the most popular.

However, from mid-May to mid-October several thousand of them take a break from soaking and hiking to sign up for a coach tour of the talc mine. Organised by the tourist office and narrated in French, this tour allows the visitors a glimpse of some of the big machines the work at the Trimouns site.

The trips occur five times a day during the peak holiday season and each hour-long tour starts at the bottom of the mountain at the local La Pierre Blanche hotel, 15 km from Luzenac. The mine entrance lies 17 km further up the winding mountain road, where visitors have the opportunity to buy books, posters, cards and even sculptures carved out of talc. Higher up the mountain, viewing points provide visitors with a spectacular glimpse of the amphitheatre of white stone. They are told the overlying rock hides a 60 to 100 m deep talc vein that extends for at least 2 km. This vein was formed over 100 million years ago so there will be sufficient reserves for decades.

The exploitation of this site started in the early nineteenth century on a small scale. The soft, white

10.8. Every summer, thousands of tourists join one of the tours organised by the local tourist office. During the tour, they are afforded stunning panoramic views of the world's largest talc mine and a glimpse of some of the big machines.

stone was ground in flour mills and the talcum powder sold to chemists in Toulouse. The development of paper mills really kick-started industrialisation at the Trimouns deposit. In 1905 the original site was called Talc de Luzenac, but later acquisition by Rio Tinto in 1988 resulted in a name change to Rio Tinto Minerals Luzenac Operations.

The mine is open for just eight months of the year. When the first snow arrives the machines are forced to retreat. As well as eleven different quality grades and colours, the talc vein at Trimouns also contains levels of chlorite – a hydrated magnesium and aluminium silicate. While this substance looks very similar to talc, it has different properties. For example, chlorite is slightly less hydrophobic (water-repellent), which makes it useful for applications that use water, such as paper making. Only skilful operators are able to distinguish between all the different colours and grades.

The obvious similarity in colour is one of the reasons why the mine cannot extract while snow is on the ground. The team has to take full advantage of the good weather conditions from April to November because they need to extract 100,000 to 150,000 tonnes to keep

the plant running during the summer. Not only that but they also need to stockpile another 300,000 tonnes for the winter months.

Big Liebherr Duo

When the conditions allow, the full-time workforce of 270 is boosted with an additional sixty-five seasonal workers. The employees work two shifts in order to extract 3,400 tonnes of talc a day. Liebherr machines feature prominently in Trimouns' machine fleet of fifty-four. To be precise, there are nineteen Liebherrs at the mine, the largest of which is a seven-year-old R 994B.

The job of this large excavator is to extract and load overburden. Seven tonnes of material have to be removed just to get at a single tonne of talc. In one season as much as 1.3 million cu m of overburden will be removed.

The mine also has a much older Liebherr R 994 that was bought new in 1987. After 35,000 hours of operation they can no longer rely on it as a prime mover. Currently used as a reserve machine, the R 994 is up for sale and has already been superseded by a new

10.9. Liebherr is the preferred choice for talc extraction. Trimouns has ten 900s, one 922, one 924 and one 934. Every day their combined efforts shift 3,400 tonnes of talc.

10.10. The Liebherr R 994B Litronic is the biggest excavator at the Trimouns talc mine. During two shifts per day, the 15.3 cu m backhoe bucket extracts over 21,000 tonnes of overburden.

10.11. There are a handful of R 994Bs still working in France.
This seven-year-old excavator has no problems filling the mine's Komatsu haulers in four passes.

machine. Its replacement is not an R 9250, as might have been expected, but a brand-new Caterpillar 992G wheeled loader instead.

The management at Trimouns prefers the mobility of a high-output wheeled loader as the second overburden machine. Overburden is stripped into 15 m sections and brought to one of two dumps on site; this involves a typical haul distance of 2 km down the mountain from the existing pit floor, which is located at approximately 1,650 m.

At this French mine, blasting occurs at the end of every working day. There is no siren to warn of an impending blast; instead Trimouns uses the century-old tradition of a cattle horn! Blown from a bench at the top of the mine, it is extremely effective and can be heard for miles.

After blasting away the overburden, and once the talc vein has been exposed, using finesse rather than brute force is the only way to remove large quantities of talc each day. Once again, Liebherr is also the preferred choice for the talc extraction excavators. They have ten front-shovel 900s, all of which are equipped with narrow booms to improve the operator's view of the talc seam face. Small shovels of just 0.80 cu m allow operators to pick and choose between the different qualities and colours of the substance. Although some of

their Liebherr excavators also carry a bigger 1.5 cu m bucket, are used mainly for loading.

Talc becomes very sticky when wet and the mine had to build a man-made track to ensure that the trucks were able to reach the loading excavators. The slabs used to build the tracks are made of reinforced concrete that is nearly 200 mm thick. The mine has enough of these 3 tonne blocks to lay over 4 km of temporary roads. Although they take time to move and set up, these tracks are crucial: a point perhaps best illustrated by the fact that their R 994B almost sank in wet talc a few years ago!

The talc is taken to a storage area, capable of holding 250,000 tonnes, where different grades are then fed into the primary crusher. Once it has been crushed, the talc is then ready for further refining so it is brought down the mountain to the processing plant at Luzenac, 1,000 m below. The crushed substance is transported not by trucks but by a bucket and cable system, making the Trimouns mine one of the few remaining global operations to rely on this technique.

The system was first installed in 1903, long before the first ski lifts appeared in the Pyrenees. The current cableway, which dates from 1976, is equipped with just

10.12. After more than 35,000 hours and twenty-two years of operation, this Liebherr R 994 is up for sale.
It is possibly one of the last excavators to wear the famous Luzenac livery.

10.13. The bucket and cable system used to transport the talc to the processing plant is unique.
Each of the 103 buckets suspended from the 5 km long cable carries 1.4 tonnes of talc down the mountain.

10.14. The mine has four of these Komatsu HD985 dump trucks which are powered by a 1,000 hp engine and have a capacity of 105 tonnes.

over a hundred steel buckets, each with a capacity of 1.4 tonnes. Loaded with crushed talc, it takes each bucket just twenty minutes to travel 5 km from the crusher to the refining site. With a maximum transport speed of 4 metres per second, the cable has a maximum capacity of 180 tonnes per hour and at full capacity a bucket is unloaded every twenty-seven seconds.

The top of the valley is prone to fierce mountain winds. The bucket and cable system is set to automatically shut down when wind speeds rise above 12 metres per second or when sensors, located 5 km to both sides, detect an approaching storm. Moving talc down to the valley in this way eliminates at least twenty road trucks, which is just as well because the narrow mountain road does not provide much room to manoeuvre.

Talc is processed into as many as forty different products and at least 150 different packaging formats. Every day up to eighty vehicles will journey with the substance to various destinations in Europe and sometimes even further afield. Like many other surface mines, the Trimouns facility is under pressure to cut costs. In an attempt to increase production rates during the short eight-month season, they have recently updated their truck fleet. With the exception of one 50 tonne hauler that was retained as a water truck, the

rest of the older machines have been replaced by two Cat 777Cs, each with a capacity of 91 tonnes, and four Komatsu HD985s, with capacities of 105 tonnes.

Cost cutting is also the reason why the new Cat 992G wheeled loader is finished in the familiar yellow-and-black livery of Caterpillar rather than the striking Luzenac livery and logo. Current mine management argue that the remote location means that there is little promotional value in finishing new machines in the mine colours; the way things are going, the Luzenac logo may be doomed to the history books.

Cat 993K

Wheeled loaders weighing one hundred tonnes, like the Caterpillar 992, are not frequently seen in Europe. In France, there are approximately twenty of these massive machines. However, the biggest wheeled loader in the country is actually the Caterpillar 993K, which weighs 137 tonnes. Though it was launched a few years ago, there are only a handful in Europe today: the first two were sold to French mines.

Eldorado Gold in Turkey has a 993K, as does Inguletsky GOK in Ukraine and there are numerous 993Ks in Africa, but Clarté quarry was the first place in Europe to purchase the massive wheeled loader. Located

10.15. This Cat 992G is the first new machine in many years not to wear Luzenac colour scheme and logo.
The 12 cu m bucket is easily capable of handling 990 tonnes of material an hour.

on the outskirts of Herbignac near Nantes, this quarry is the largest of seven operated by the family-owned Charier CM business. Clarté is typical of many aggregate quarries in the Atlantic Coastal region in Brittany. While sales of construction machines have plummeted during recent years, aggregate demand remains at a high level. Many quarries are not only trying to meet existing production targets but also planning for future increases.

This is the problem faced by Charier. The quarry was founded more than a century ago and its preference for the Caterpillar brand began in 1939, over seventy years ago, with the delivery of a D4 dozer. In 2001, a Cat 992G wheeled loader became the quarry's prime mover. On a good day, it was capable of filling an average of ninety Cat 777Ds with 9,000 tonnes of material – meeting annual targets of 1.9 million tonnes. However, after 17,000 hours of intense operation, its best days were over. The obvious replacement candidate was a brand-new 992K, but the quarry wanted to raise outputs to 2.5 million tonnes a year and they were not sure that the 992K would be up to the task of doing so all on its own.

French Caterpillar dealer Bergerat Monnoyeur was consulted and they calculated that a smaller 988 would be perfectly adequate for their requirements but it would have to work two 8-hour shifts per day, instead of just the one shift. The quarry was not keen on this suggestion because of the fear that this would lead to a considerable increase in production costs. Adamant that whatever machine was purchased, they would stick to a single 8-hour shift per day, the management at Clarté quarry soon concluded that what they really needed was a bit more power. While the 992K, the successor of the 992G, boasted a host of new features, one thing it lacked was the ability to provide much more production output.

The problem was that the next model in Cat's wheeled loader family was the 994F, one of the world's largest mechanical versions. The enormous machine weighed an incredible 200 tonnes, too big for Clarté. The gap between the 992 model and the 994 model was simply too great and it seemed like the French quarry would be stuck. Then Caterpillar brought out the 993K and Clarté became the very first European customer.

10.16. With an operating weight of 137 tonnes, the 993K is the second largest wheeled loader made by Caterpillar. This machine was made in the US and was the very first 993K commissioned in Europe.

The 993K premiered in September 2007 and promptly went into production the next year at the Caterpillar factory in Aurora, in the US state of Illinois. The Aurora factory also makes excavators, wheeled dozers and four other massive wheeled loaders – the 988H, 990H, 992K and 994F. The quarry was quickly convinced that the 993K was exactly what they were looking for and the parts of these movers soon arrived at Le Havre. The French quarry bought the 993K for between US$2.3 and 2.5 million and it was the first in Europe.

Just as the extra 30 tonnes of operating weight would suggest, the 993K is a good deal heavier than the quarry's old 992G – thirty-one per cent heavier to be precise. Physically it is also a much larger beast. Standing 6 m tall, from the ground to top of the cab, the 993K stretches to a length of over 15 m. Four vehicles were needed to shift the components from Le Havre to the quarry. Then it took a further two weeks – with the help of a crane, two mechanics and four of the Caterpillar dealer staff – to prepare the giant loader for action.

Though both the 993K and the 992K are equally new, they have little in common. For example, unlike the 992K, the 993K has a new wheel size, engine, lift arms and bucket. However, all Caterpillar machines do share certain technologies. These can include the service

brakes, which are similar to those brake components used in the D9R dozer. Also, the planetary grouping of the transmission is similar to that of the larger 994F wheeled loader and the final drives are built from the design platform of the D11 dozer, but then tailored for a wheeled loader application.

The major difference between 993K and the 992K is that the 993K is certainly more powerful. Rather than taking the 3508B engine from the 992G, the newcomer runs on a powerful Tier II C32 Acert version. With a net power rating of 1,064 hp, this same 12-cylinder, 32 litre engine powers the D11T, 992K, 854K and 777F.

The new prime mover at Clarté has been responsible for boosting output by at least ten per cent. The Cat 993K has no trouble filling a hundred trucks during a single shift and is easily capable of clearing away a 30,000 tonne blast in two or three days. With a rated payload of 25 tonnes – which increased recently to 30 tonnes on new standard-lift models – the machine is capable of filling the quarry's two Cat 777 trucks in just four passes, which is one less than the 992G. In fact, Clarté claims that in the right conditions the 993K can load them in three decent passes: a real time-saver!

10.17. Each year, millions of British tourists travel through the popular French ports of Boulogne and Calais and have no idea that they are close to the largest wheeled loader in the country.

Marble Movers

The second Cat 993K in France is almost visible from the A16/E402 motorway linking the popular French ports of Calais and Boulogne-sur-Mer. Millions of tourists pass through these two towns each year totally unaware of their close proximity to the largest wheeled loader in the country. The ferocious machine works at the largest of a trio of limestone quarries that are located close to the village of Ferques. With an annual production rate that varies from 6 to 8 million tonnes, Carrières du Boulonnais is the biggest quarry in France today.

The presence of the 993K may not be widely known, but the area is famous for two things. The first is the Cap Gris Nez (Grey nose cape), which is the point of France that is closest to England and, on a good day, provides unbeatable views of the white cliffs of Dover. Second, the area is nationally famous for its marble quarrying activities. Over a century ago, huge blocks of grey/black marble were extracted from quarries surrounding the village of Ferques. The statue of Napoleon in Boulogne-sur-Mer and a number of buildings in Paris – including the impressive Gard du Nord – were carved from marble extracted from this area.

For tourists who are interested in geological history, the Maison du Marbre et de la Géologie in Rinxent is a useful information point. It provides a great opportunity to rub shoulders with some of the blocks, gain an insight into former times.

All three limestone quarries, including Carrières du

10.18. The marble museum at Rinxent is a great place to learn more about marble and organise a visit to a local quarry.

Boulonnais, owe their entire existence to the demand for marble. However, large-scale marble extraction slowed enormously towards the end of the 1960s because white and green marble became trendy and black marble went out of fashion. Ultimately, the Carrières du Boulonnais site was closed due to an inability to compete with higher quality imports from Italy and China.

Although large-scale commercial extraction ceased a decade ago, marble is still extracted by a private company near the old sites at the appropriately named Napoleon quarry.

10.19. While it is not easy to access the modern-day quarrying activities at Carrières du Boulonnais, a public viewing point provides great views into the Napoleon marble quarry.

Not far away, the site of Carrières du Boulonnais is now owned by the Poulain family and as France's largest quarry it provides work for around 300 people. The compacted limestone extracted from the site is processed into over a hundred different products, ranging from asphalt fillers to washed or dry sand and from rock armour to aggregates for railways. The location is simply ideal. The quarry is in close proximity to a motorway and to make things easier the Poulain family have their own railhead as well as a private barge-loading facility at Boulogne-sur-Mer.

Carrières du Boulonnais has two features that are not to be missed. First, the inner edges of all the haul roads are lined by an unbroken row of huge limestone blocks, each of which is plenty big enough to stop a truck dead in its tracks. This is a safety precaution to prevent a vehicle from driving over the edge of a 15 m high bench. Second, the quarry is the domain of an impressive fleet of wheeled loaders, which includes five Komatsu WA600s, each weighing 50 tonnes. A few years ago, a Komatsu WA900, weighing 100 tonnes, worked at the front line. Unfortunately, this machine has been replaced by another 100 tonne wheeled loader – a Cat 992G. Carrières du Boulonnais has also bought a 993K.

When questioned about the decision to switch from Komatsu to Cat, the quarry production manager admits his choice has to do with Caterpillar's speed of service during a breakdown. The five-year-old 992G and its much newer companion, the 993K, spend their days filling seven Cat trucks, each with a capacity of 90 tonnes. After 20,000 hours of operation, the older

haulers were slowly being updated but the recession scuppered plans to complete the change. Now it seems that quarry management are hoping that all five trucks will keep going until the economy shows signs of a proper recovery

Wielding a bucket that measures 15 cu m, the 993K is clearly meeting expectations: easily loading an average of 130 – and sometimes as many as 150 – trucks during an eight-hour shift. Together the 992G, equipped with a bucket of 12 cu m, and the 993K are barely challenged by their task of handling 2,000 tonnes an hour. In fact, during its first year at the quarry the 993K handled 4 million tonnes of material – double that of the 992G.

In the future, mine management want to boost hourly outputs to 2,500 and possibly 3,000 tonnes per hour, so the extra performance of the 993K will be a crucial asset. The mine recently carried out a small test to find out the limits of their wheeled loader. The results of this test confirmed that working close to the crusher it is easily capable of shifting 2,000 tonnes of material an hour.

When the time comes to replace the Cat 992G, the French quarry is considering swapping it for an excavator. The seams on the site contain a tough mixture of clay and limestone, which is difficult for a wheeled loader to tackle. Many years ago, an electric O&K RH120C front shovel was the prime mover; however, with more than 1.5 km of rock face, they found that it was too slow to move around.

The plan is to replace the 992G once it reaches 30,000 hours of operation. So far it has done around 25,000 hours of work, which means that mine management still

10.20. Wearing chains to reduce tyre damage, the Cat 993K at Carrières du Boulonnais is one of just two in France.

have time to decide whether to stick to wheels or change to tracks. If they do decide to buy an excavator, then it will provide the brute force needed to mix and match the different materials on the site, so that the 993K wheeled loader can pick those materials up.

The limestone at Carrières du Boulonnais contains high levels of calcium and magnesium carbonate, making it ideal for a wide range of applications, including but not limited to cement or aggregates for road building. Arcelor Mittal takes nearly one million tonnes of blast furnace sand for its Dunkirk steelworks, while a local sugar factory uses the limestone to purify the sugar, which makes it white.

The current 150 m deep pit has taken fifty years to develop and the operators have no intention of slowing down. The current extraction licence is valid until 2028, by which time they will need to have applied for another one in order to continue excavating the reserves of limestone, which are estimated at 450 million tonnes. That amount of limestone is enough to keep the quarry busy for the next ninety years and several more generations of the Poulain family will be required to finish the job.

The First Hydraulic Excavator

In 1951, the earthmoving world was changed forever with the invention of the first hydraulic excavator. Created by a French farmer called Georges Bataille, these new machines were purchased in more than 120 countries – including Japan, China and the US – and distributed under the company name of Poclain. However, when the fuel crisis hit in autumn 1973, things started to go horribly wrong for the company and they were unable to obtain any financial support from the banks or the French government.

In 1977, Case Tenneco, an American-based company, offered a rescue package which resulted in their acquisition of forty per cent of Poclain. By 1985, Case Tenneco had secured a majority holding in the French company and, four years later, the American company had acquired all the remaining Poclain shares.

Shortly thereafter, the company began trading as Case-Poclain and the machines they produced made the transition from the red-and-white Poclain livery to present-day Case colour. Despite the takeover, the Poclain name is still alive in France: operating as Poclain Hydraulics, the company manufactures hydraulic components and is still owned and operated by the Bataille family.

The site of the former Poclain factory at Le Plessis-Belleville – which is located 30 km north-east of Charles de Gaulle Airport, near Paris – is now the Espace Georges Bataille business centre. As well as housing the Case sales department, the site also provides space to store Case spare parts in a former Poclain components building as well as accommodation for nearly a hundred different companies.

Nearly 84,000 Poclain excavators were produced during the company's glory years but sadly, few survive today. While many of the old machines ended up in Africa because they could be run more cheaply there, most of those that remained in Europe were scrapped. The fact is that the high price of scrap metal was an incentive for companies to part with their old Poclain excavators.

French Fan Club

The last Poclain excavator left the production line twenty-five years ago, but the brand still has a huge following. Association Génération Deux is a French organisation created with the sole aim to preserve the heritage of the Poclain brand. The organisation is open to all former Poclain employees and friends. The vice-president of Génération Deux, Philippe Fritz, worked for the French excavator manufacturer for thirty-five years and is not only an invaluable source of knowledge but also the driving force behind the organisation. After moving from the technical department to marketing, he was then appointed product manager for heavy excavators and finally became the excavator product director.

Philippe is pleased with the interest that Génération Deux has generated. He confirms they now have around 300 members. 'The majority of our members are from France, Belgium, and Germany,' he says, 'but some are from as far away as Brazil and New Caledonia.'

For the past ten years members of the organisation have worked hard to recover files, microfilmed drawings, former Poclain advertisements and engineering models. The hope is that in doing so they will prevent Poclain memorabilia from being lost forever. Génération Deux currently has numerous pictures and brochures uploaded onto their informative website, which is in French. Future plans for the group include developing a Poclain foundation and building a museum to exhibit a range of Poclain machines.

Poclain Junkies

Elsewhere in France, three enthusiasts are also doing their very best to secure the future of as many Poclain excavators as possible. On the outskirts of the village of Saint-Pierre-de-Boeuf, is the biggest single collection of Poclain excavators in Europe. Owned by Eric Moutot, an enthusiastic member of Génération Deux, these Poclain excavators are in varying states of disrepair. The uninspiring site where they are housed contains approximately thirty excavators: some are barely recognisable, while others have been painstakingly restored to their former glory by Moutot and his equally fanatic colleagues Fabrice Santoro and Sylvain Colasseau. The men call themselves 'Poclain junkies' and admit that they would not be able to live without the brand.

Moutot supports his habit with his business. During the past twenty years, he has created a thriving contracting business which disposes of 100,000 tonnes of hardened leftovers from local cement plants. Contracted by many of the big names, such as Holcim, Cemex and Lafarge, Moutot's successful operation covers a 200 km radius, which includes Grenoble, Lyon and Valence. He relies on a fleet of four present-day Case excavators to execute this tough work.

Spare time is precious to Moutot and what little time he does have is devoted to the collection and restoration of old Poclain machines. Moutot, Santoro and Colasseau have a simple mission: to collect virtually every Poclain model ever built – both wheeled and tracked. Their unique collection already includes a well-restored, 60 tonne 350CK – quite possibly the only remaining one in the world that works. This machine is kept at a quarry near Lyon. While none of their restored excavators will ever see any real work, occasionally they try and find a site to play with them!

At their yard is another completed restoration project – a 610CK – which is a Poclain heavy-duty crawler. The machine was originally launched as the 600CK, but due to complaints from the 600 Financial Group Poclain later changed it to the 610CK. While the name might suggest that this excavator weighed around 60 tonnes, this particular model actually weighed 120 tonnes: a testament to the sturdy build of the Poclain excavator.

Poclain 1000CK

The machine that gets the most attention at Moutot's yard is the 1000CK. It is not only the rarest excavator ever made by Poclain but also the biggest. In fact, when the 1000CK was first brought out it was easily the

10.21. The three men have done a great job with this 610CK which they bought from a coal mine in Clermond Ferrand. All the hand-made sheet metal needed to be replaced but the excavator still has many original features.

largest hydraulic excavator in the world with an operating weight capable of varying from just under 150 tonnes to a maximum of 210 tonnes. During the period of 1975 to 1983, Poclain made just fifty-seven of these massive machines – thirteen of which were purchased by operators in the UK. Unfortunately, it is unlikely that any survive today.

Poclain actually built three versions of their biggest-ever excavator. The first was the EC 1000, which was launched in April 1970. It was powered by three GM engines, which produced 280 hp each. The machine had an operating weight of 147 tonnes. Research suggests that only six models of this kind were produced and none of them exist today.

The second version was the 1000CK M1. Eric Moutot owns this version, which he purchased from a Spanish contractor called Dragados; it was one of only ten made in 1978 by Poclain. In its heyday, this 1000CK used to work alongside a second identical machine at the Puentes de Garcia Rodriguez coal mine, which is located not far from La Coruña in the north-west of Spain. These two heavy-duty excavators were capable of

shifting an average of 1,000 cu m of material per hour.

Moutot still has all the component parts for his excavator on site. When the 9.6 m boom, 4.5 m stick and 6.8 cu m backhoe bucket are reunited the entire weight of the machine will be around 157 tonnes. This second version featured a single-seat cab and two Deutz V12 engines capable of producing a maximum of 792 hp.

There is no doubt that Eric Moutot has a very rare machine. However, there is one other 1000CK in France; this model is the 1000CK M2 and it currently resides with a collector in the north-east of the country.

Originally launched in 1979, this third model boasts a new upper structure, featuring a cab with two seats, as well as Cummins engines and a bottom dump shovel attachment. The French collectors' renovated M2 model has a boom measuring 5.5 m, a stick measuring 4 m and 10 cu m standard bucket. The weight of this model varied according to options. For example, if the excavator came with track pads measuring 1.1 m wide, then it would weigh approximately 191 tonnes. On the other hand, if the M2 was purchased with a complete set of options, then weight increased to 210 tonnes.

10.22. With an operating weight of 190 tonnes and a maximum output of 792 hp, the Poclain 1000CK was the biggest excavator ever made by the French manufacturer. Eric Moutot's excavator is believed to be one of the just two Poclain 1000CK excavators in France.

10.23. Launched in 1970, this 147 tonne EC 1000 was powered by three GM 280 hp engines.
Only six machines were made and none of them exist today.

10.24. As well as three EC 1000s British mining company Miller used to have a single 1000CK M1.

Though they have the biggest Poclain excavator ever built in their collection, Eric Moutot and his colleagues are keen to expand further. In the near future, they would like to acquire a 90CKB and a 61CK but in the meantime, they are not short of things to do. The main priority at the moment is the construction of a new workshop and a concrete area to house the restored models. As for the machines, though there are plenty of excavators on site, the 1000CK attracts all the attention

so they hope to complete it within the next couple of years. Once it is finished, they are likely to be swamped by requests to view it in action.

One thing is certain, even without adding any more models they will need at least twenty years to restore what they currently have. Excavation restoration projects are also very expensive, making this not just a question of time but also of cost.

10.25. The cab of the 1000CK M2 was widened to seat two – the operator and the grease monkey.

10.26. This dilapidated 400CK was sourced from a Belgian limestone quarry. Moutot and colleagues plan to restore it to mint condition, which could take quite a while.

CHAPTER 11

Norway

The landscape of Norway is quite simply astounding. Retreating glaciers have carved up huge expanses of land and left the valley bottoms along the rugged west coast littered with many millions of boulders, some the size of houses. Fierce winds blowing in from the North Sea have rounded many mountain peaks in the country, but their average height continues to increase at the rate of several millimetres a year due to tectonic activity.

Each year a handful of aggregates quarries carve out nearly a million tonnes of stone from the jagged mountainsides. However, the efforts of these smaller operations are completely dwarfed by the Jelsa quarry, which is jointly owned by Norsk Stein (Norwegian Stone) and the Heidelberg Group. Jelsa is not just the biggest quarry in Norway; it is the biggest one in Scandinavia. The introduction of recent expansion measures has resulted in the quarry growing even larger, with production being pushed to 3,000 tonnes per hour. A big chunk of the money – approximately €113 million – that was earmarked for the expansion programme was spent on a brand-new fine Metso crushing and screening system. The installation of this high-tech equipment, which came from Egypt, has elevated Jelsa's output to industry-leading levels in Europe.

With 400 million tonnes of high-quality reserves, the quarry confidently expects to churn out 15 million tonnes of aggregate a year for the next forty years. It is located on the edge of the Ryfylke Fjord, and although Jelsa is roughly 100 km north-east of Stavanger, it can take three to four hours to get there from Stavanger Airport. Gauging road distances in this part of the country is not easy; even though more new tunnels dip deep under the fjords, many roads still end abruptly at a ferry crossing.

11.1. The Jelsa quarry lies on the edge of the Ryfylke Fjord. In this unforgiving climate it can rain for weeks at a time.

11.2. Almost all 15 million tonnes of aggregates produced at the quarry leave by boat.
Jelsa is one of the largest coastal aggregates quarries in the world.

Difficult as the quarry is to reach by road its stunning location affords direct links to the North Sea. The quarry makes use of this convenient transportation link by loading almost all of its aggregates onto boats. Some of the thousand vessels that dock at ports in this area each year are capable of carrying up to 100,000 tonnes of material. Destined mainly for France, the Benelux region, Germany, Denmark and the UK, their high-quality aggregates are popular for road building, railways and offshore purposes.

Today, Jelsa is one of the largest coastal aggregates quarries in the world and it accounts for fifty per cent of Norway's total aggregates exports. This has not always been the case, however, since Jelsa was actually born out of humble beginnings. In the mid 1980s, following the completion of a local hydroelectric dam project, two of the employees purchased the crusher used in the scheme and moved it to the current site to sell crushed stone.

The employees at Pon Equipment, the Norwegian Caterpillar dealer, still recall the very first machine that was sold to the quarry in 1987. That machine, the 966E wheeled loader, remained the prime mover at Jelsa until 1990. The quarry then went through a spell of alternating between wheeled loaders and face-shovel excavators. The first excavator was a Norwegian-made Brøyt X52, weighing 55 tonnes, which was eventually replaced by a Cat 992 wheeled loader. Three years later, the wheeled loader was swapped for a Hitachi EX1800 – a large hydraulic excavator. This excavator did not last long at the Jelsa site, however, because quarry management found the running costs a bit high. The powerful digger was shipped off to Canada after only five years. It was replaced by three Cat 988H wheeled loaders and an even bigger Cat 992G.

The Cat 992G was officially the quarry's prime mover and one of just a dozen sold in Norway over the past ten years. The bucket attached to this machine measured 11.5 cu m and was capable of lifting nearly 25 tonnes of material at a time. With 10 tonnes of tyre chains attached, the Cat 992G tipped the scales at around 110 tonnes.

The reign of the Cat 992G was a short one, however. The expansion project at Jelsa brought about a wind of change through the machine fleet and the Cat wheeled loaders are no longer the prime movers of choice at the quarry. They have, in fact, been ousted in favour of four Komatsu versions – two WA800s and two WA900s – each weighing around 100 tonnes. However, these were

11.3. They are planning to swap three Komatsu HD785-6 trucks for new Dash 7 versions. Each one carries 100 tonnes.

11.4. One of two Komatsu WA800 wheeled loaders at Jelsa.
They also have a pair of WA900s and it has been hinted that management may invest in more power in the future.

not the only new machines supplied by Norwegian Komatsu dealer Hesselberg Maskin: the quarry also took two PC6000 excavators, each weighing 60 tonnes.

The quarry has a policy of replacing its machines after 6,000 hours – equal to roughly two years of operation. Management at the quarry think that wheeled loaders provide trouble-free operation within this time frame and a benefit of owning new machines is that it makes it easier to find willing operators. This is crucial given the quarry's remote location and that fact that the current expansion programme requires many more operators to facilitate growth.

The burning question now is whether the four Komatsu loaders are up to the task of meeting the ambitious projected outputs of 15 million tonnes a year, or whether even more muscle will need to be brought in. While there is a hint of even more power, for now Norway's largest wheeled loaders work further south.

Titania

The tiny village of Sogndalstrand is famous in Norway for its well-preserved seventeenth- and eighteenth-

century wooden structures and its stunning views out to sea. This coastal gem is located two hours south of Stavanger in the community of Sokndal. Further along the coast is the town of Jøssingfjord, the site of the incident between marines from the British destroyer *HMS Cossack* and the German tanker *Altmark* back in 1940, and Rekefjord, home to the Fjordstein aggregates quarry.

Despite the popularity of this coastal region, many tourists never even notice the overhead conveyor, which feeds boats with ilmenite concentrate extracted from a mine higher up the mountain. The Tellnes deposit and surrounding 18 sq km of land is mined by Titania A/S, which is owned by Dallas-based Kronos Worldwide Incorporated. This Norwegian mine produces six per cent of the world's ilmenite requirements, making it not only the largest known deposit found in Europe, but in the entire planet.

In 2008, Tellnes shot to fame when Katie Melua agreed to perform a one-off summer concert at the site. The bottom of the void had to be drained of water and specially levelled to receive 6,000 visitors for the event, the massive excavators and trucks forced to make way for the large number of coaches providing transport to the location.

11.5. The Tellnes deposit yields six per cent of total global ilmenite production, making it the largest producer in the world.

11.6. *The bottom of the pit had to be drained and the big machines forced to retire when Tellnes played host to 6,000 visitors eager to see Katie Melua in concert.*

Though Tellnes may have only recently come to the attention of many, the site has actually been in operation approximately fifty years. While more than 200 million tonnes of ore and waste have been extracted during that time, it is estimated that below the surface there are still another 200 million tonnes of reserves: a sufficient amount to keep the present workforce of 250 employees fully occupied for the next sixty-five years.

The first ilmenite was discovered in the area back in 1902 at an underground location not far from the current operation. Mining in this first location began in 1916 but by 1954 supplies were seriously depleted and a new deposit was needed. A thorough aerial survey of the area revealed the Tellnes deposit, which was fully operational just six years later.

Black ilmenite is the result of a volcanic intrusion. At the site, this material is crushed and shipped to sister companies in Norway and Germany, as well as to customers in Finland, Poland and the Czech Republic. When processed further the ilmenite becomes a white titanium dioxide pigment, which is used as a brightening agent in paints, plastics, paper and cosmetics.

Surveys show that the ilmenite is lying at a forty-five degree angle within the sediment. Mine management admit that they are unable to estimate the exact size of the deposit at Tellnes but they are not worried because even when surface mining is exhausted they will still have the option to go underground. In the past, there was an underground operation in place but it was abandoned in 1965. Despite this, the deposit is still thought to contain an estimated 60 million tonnes of ilmenite. In the future, mine officials might consider creating a tunnel to allow them to tap into this substantial supply but for now Tellnes is busy with the surface operation.

A few years ago, ilmenite output was at 750,000 tonnes a year, and with soaring profit levels there was renewed confidence in expansion so work began on extending the void to boost production further. To this day, work on the push-back continues but Titania has not escaped the global economic downturn. Recent plans to extract between 850,000 and 1 million tonnes of ilmenite a year have had to be revised: production has actually been trimmed by twenty per cent to around 760,000 tonnes per year.

11.7. This machine is the older of two O&K RH120Cs that have performed twenty years of hard work. It was retired a few years ago so that its vital organs could be stripped for the second machine.

Many of the mine's workforce live in the local town of Haug i Dalane and Arvid Johnsen is one of them. He has worked at Tellnes for thirty years. After starting as a drill operator, he sat on a wide variety of machines including graders, trucks and wheeled loaders. Arvid also spent a four-year stint on a P&H rope shovel.

His wife and eldest daughter also work at the mine, the former works in the processing plant and the latter driving a truck. Arvid has seen many changes, but one thing that has never altered is the mine's heavy reliance on the largest earthmovers in the country. At the turn of the century, Tellnes had a 365 tonne P&H 1900 rope shovel and it was one of the largest of five prime movers at the site. The others were also unique machines in Norway. Tellnes laid claim to two O&K RH120C excavators, each weighing 213 tonnes, and two LeTourneau wheeled loaders: one L-1000 that weighed 100 tonnes with an 11 cu m bucket and one L-1100 that weighed 120 tonnes with a 14 cu m bucket.

The rope shovel is now long gone and one of the O&K RH120Cs was retired in 2007 after operating for an impressive 62,700 hours and handling 26 million tonnes of material. Exhausted after twenty years of punishing work, the O&K machine was stripped of its vital organs

to keep the second machine running. The other O&K RH120C was built in 1989 and was fortunate enough to have been recently overhauled; however, its days are now numbered as well.

The future of the two LeTourneau machines is also no longer safe. During a twenty-year reign, the 1,200 hp LeTourneau L-1000 has worked an astonishing 55,000 hours without the slightest bit of trouble. Now relieved of its role as prime mover, the machine is on sale to the highest bidder. Until the management at Tellnes find a buyer, the L-1000 will spend its time doing relatively easy work near the main crusher.

The ilmenite vein at Tellnes contains different qualities of the substance, all of which are blended at the crusher. Ore from one particular area of the vein is stockpiled near the crusher and fed, at intervals, into the machine by the wheeled loader. The truck contents from another part of the main ore body are tipped directly into the crusher. Handling up to 1,200 tonnes per hour, the crusher eats 12,000 tonnes per day in order to meet current production targets.

The second LeTourneau loader – the 1,400 hp L-1100 – has chalked up 30,000 hours, and is still very much active. Capable of lifting 25 tonnes with apparent ease,

11.8. Built in 1989, the second O&K RH120C was still used to load ilmenite a couple years ago.
The drop in production means that even this machine is now on standby and could soon face the gas torch.

11.9. This LeTourneau L-1100 is another machine that has amassed many thousands of hours of operation
– more than 30,000 to be precise. It might be old but it still lifts 25 tonnes with ease.

11.10. In a good blast the L-1100 can fill a truck with a 180 tonne capacity in just seven passes. It is another twenty-year-old machine that has given over 55,000 hours of operation.

the L-1100 fills their 180 tonne capacity trucks in just seven attempts after a good blast. Unfortunately, mine management has mixed feelings on the two American-made loaders. They were delivered new in 2000 and the mine experienced a number of problems with the L-1100 during the first two years of operation. In a good year the LeTourneau can handle 2 million tonnes of material but there have been times when the machine has been out of action for six weeks at a time with all manner of problems. If the mine is going to continue to enlarge the void at Tellnes, then they will need more reliability and power. The emphasis is not just on renewal of current machines but also on the reduction of the number of large excavators on site.

Powerful God

Evidence of Titania's prime-mover renewal programme started with the commissioning of a brand-new Terex O&K excavator to replace the older RH120C. The RH170 is, by European standards, a seriously large machine. Made in a factory in Dortmund, Germany, the

374 tonne machine is easily the biggest hydraulic excavator in Norway and one of only a few currently operating in Europe. The machine has been nicknamed Tyr by the staff of Tellnes, after the mythical god of victory and glory who is portrayed as only having one hand.

The excavator, though it is a Terex, was sold to the mine by Pon Equipment, Caterpillar's Scandinavian dealer. Caterpillar has big plans to return to the mining excavator market, but when it stopped making its 5000 series it left its dealers – Pon Equipment included – with a gaping hole in their product range. This problem was resolved following a marketing deal which gave many Cat dealers access to selling large Terex mining excavators. The Norwegian Pon sales team were so excited that they claimed they would take a dip in the North Sea if they could succeed in securing a sale of this first machine. True to their word, with temperatures of just five degrees, a number of employees took the plunge after management at Tellnes agreed to purchase the O&K RH170!

The journey for Norway's largest excavator was labour-intensive but fairly straightforward. Eleven vehicles were

11.11. After having a good experience with a pair of O&K RH120Cs, Titania decided to invest in a Terex O&K RH170. This machine is easily the largest hydraulic excavator in the country.

needed to shift the hefty components of the O&K RH170 from the factory at Dortmund to Bremen in northern Germany. From there the parts went by boat to Egersund in Norway and the machine was assembled and ready for action ten days later.

Although Norway's largest excavator was expected to work seventeen shifts a week – equal to between 5,000 and 6,000 hours of active duty a year – things did not start well. Electrical faults were commonplace and there were problems with the automatic lubrication system to the front shovel. As a result, during the first three years at Tellnes the O&K RH170 worked for a total of just 10,000 hours.

11.12. An RH170 is a rare sight in Europe. With an operating weight of 374 tonnes, Titania's RH170 is dubbed Tyr, after a mythical Norwegian god. It is certainly powerful because the twin engines pump out 2,000 hp.

Despite this slow start, the performance of this massive excavator is now meeting the expectations of mine management. The RH170 is well-matched to the mine's 180 tonne capacity trucks and can fill them in just five passes. The excavator came with an 18 cu m bucket – capable of lifting 35 to 40 tonnes at a time – but this has since been swapped for an even larger 20 cu m one and the excavator upgraded with a newer B series attachment.

Terex developed a new boom and stick design for the RH170B featuring a stronger, but lighter, internal framework. This allows mechanics to fit a bucket that is 2 cu m larger than the original. The new bucket is capable of shifting 40

11.13. Black ilmenite (left) is quite distinctive. When ground into a white titanium dioxide pigment, it is used as a brightening agent in paints, plastics and cosmetics.

to 45 tonnes of material per fill. The power of the big O&K RH170 comes from two 12-cylinder Cummins engines, the combined efforts of which pump out a maximum of 2,000 hp. A big excavator needs a big fuel tank and the one on this machine can carry up to 6,300 litres of diesel.

It is just as well that the Terex is an enormous and powerful machine because if the Tellnes expansion project continues there will be serious amounts of material to move. Even in the main pit almost 3 tonnes of overburden have to be removed in order to get to just 1 tonne of ilmenite concentrate. This ratio rises to 7/8:1 in the extension.

Moving Mountains

There is so much overburden to shift at Titania's Tellnes mine that the Terex O&K RH170 cannot manage the task alone. To help keep up production levels another massive machine has joined the team – a Komatsu WA1200. This machine is the largest mechanical wheeled loader in the world and it is one of only two found in Europe. The first is discussed in Chapter 4 on Finland.

Almost twice the size of the LeTourneau excavators, this high-capacity wheeled loader was purchased because

overburden production will rise significantly when the new Tellnes extension opens. Bought to replace the older LeTourneau L-1000, the parts of the Komatsu WA1200 were first shipped to the port of Egersund and then transported to the mine on the back of twenty vehicles. One month after the arrival of the parts the 205 tonne giant was ready for action. With chains fitted to all four massive tyres, each one weighs 5 tonnes, meaning that the overall operating weight of the WA1200 is 225 tonnes. Loading 40 to 45 tonnes at a time, the ground trembles as the powerful machine forces its way into the heap. It keeps three new Cat 789C trucks fully employed, requiring just four passes to fill each.

Titania has always boasted having the biggest trucks in Norway. In 1984 it bought a 136 tonne capacity Komatsu HD1200. Eight years later it added the first Unit Rig MT-3600 in Norway and in 1995 followed that with a second one. Both trucks had a payload of 172 tonnes but 60,000 hours later they are both now in need of replacements.

Mine management are pleased with the performance of a 177 tonne Cat 789B that was purchased a few years ago and so the ageing Unit Rigs are slowly being axed for brand-new Cat 789Cs. The availability of these new machines is good and each one is capable of carrying around 1.5 million tonnes of material a year.

11.14. The WA1200 at Titania is only the second one in Europe. The 20 cu m bucket loads 40 to 45 tonnes at a time, easily filling Cat 789C trucks in just four attempts.

11.15. This Komatsu WA1200 has an operating weight of 205 tonnes and the four tyre chains boost the weight of this phenomenal machine by another 20 tonnes.

11.16. The mine has always boasted owning the biggest dump trucks in Norway. Back in the 1990s it had the only two Unit Rig MT-3600s in the country and now it has 177 tonne capacity Cat 789s – once again the largest trucks in the country.

One of them was recently in the workshop for a tipping body modification. With extra length and height added to the machine, the carrying capacity of the Cat 789C is boosted from 105 to 120 cu m. The reason for this is simple: the density of the overburden is a good deal lower than ilmenite so at 3.4 tonnes per cu m the ore is heavy, whereas the overburden is much lighter. Before the modification, a truck that was fully loaded with overburden could carry just 130 tonnes. This was not enough for the mine so the modifications will seek to improve that carrying capacity. As it stands, two of the truck bodies have already been modified and the other two will follow shortly.

Early Warning

Finally, health and safety is a big issue for mines located all over Europe. The concerns are not limited to machinery but also include the mine conditions and the surrounding area. At Tellnes, one side of the void is unstable and rock slides are commonplace. This risk increases when it rains, an unavoidable factor in Norway. Mine management are currently evaluating an early warning system called the Slope Stability Radar (SSR). This system is one of sixty systems installed globally but it would be the first in Europe. SSR was created by the Australian company Ground Probe and it is billed as the only device that provides continuous sub-millimetre measurements of rock movements across the entire face of a wall. Supported from South Africa, it allows geotechnical engineers and mine personnel to track movements confidently. Taking just ten to twelve minutes to scan the entire side of the mine, it provides an advanced warning of any size of rock movement, from just a few tonnes to several million.

Currently, mine managers can be alerted to any movement through their mobile phones but it is only a matter of time before all staff will be equipped to receive warning text messages. This provides sufficient time for the removal of equipment and people from risk areas. So

11.17. Despite even the best safety systems, surface mining is a difficult and dangerous business.
This Slope Stability Radar (SSR) is an early warning system.

far, they have only been forced to evacuate once when a 70 m section moved 50 mm in just a few hours. 'We quickly pulled all men and machines out of the area,' says mine manager Knut Petter Netland. The system is definitely not cheap but as the alternative could be much worse, he is quite certain that it is well worth the investment. 'We have to do everything we can to prevent accidents,' he says. 'This system could help to reduce the risks.'

CHAPTER 12

Belgium

Belgium, one of the smallest countries in Europe, is best known in earthmoving circles as the home of Caterpillar's largest global factory. The hydraulic excavators and wheeled loaders made at the factory in Gosselies are transported all over the world but some of them work in the scores of limestone quarries which stretch right across Wallonia – the French-speaking, southern half of the country.

Limestone's importance to this region spans many centuries so it is not surprising that evidence of former quarrying activities is easy to find, with some exhausted quarries transformed into tourist attractions. One of the best known of these transformed quarries is Dinant Adventure, a sports paradise overlooking the river Maas.

South of Brussels, water is currently being drained from another old quarry at Lessen as part of an €80 million plan to transform the land into the largest indoor ski centre in the world. The centre will be called

Snow Games when finished and nearly 5 million cu m of artificial snow will be needed to cover the 10 hectares of covered slopes, ready to receive an estimated million visitors a year.

Still further south, toward the French border, some older workings have been given a new lease of life as the Aqua-Tournai leisure centre, just 3 km from Tournai city centre. Limestone has been excavated around the 2,000-year-old city since Roman times, when it functioned as a stopover on the road from Cologne, Germany to Boulogne on the French Coast. The material extracted from this area is sometimes incorrectly referred to as Tournai marble and was originally carved out by hand and used for building.

The dark limestone was laid down in the carboniferous period about 350 million years ago. Hard enough to make many of the town's older buildings,

12.1. Limestone has been extracted for many centuries near Tournai and still many of the big names in the Belgian limestone world – such as CBR and Holcim – run quarries just a few kilometres from the city.

12.2. Much of the aggregates and clinker from this region is shipped along the river Scheldt (Escaut) –
a 350 km long river that originates in France and ends in the North Sea. Some of the aggregates extracted from
Carrières d'Antoing can be seen at the rear right of this picture.

pavements and kerb-stones, the limestone began to be extracted by full-scale industrial processes in the early 1900s. At that time, the town was surrounded by nearly a hundred quarries: the reason for popularity was obvious – there were heaps of limestone and the seams were all found close to the surface.

Between World War I and World War II the industry consolidated and quarry numbers declined to just ten large sites, all of which were kept busy supplying cement and aggregates to help expand the Belgian road network. Today, quite a few Belgian limestone quarries produce upwards of half a million tonnes a year.

However, two of the four industry-leading operations that flank the banks of the River Escaut (Scheldt) each produce 2 million tonnes a year, while the largest – Carrières d'Antoing – churns out twice that amount.

Carrières d'Antoing is located just outside Antoing, a town with a population of around 7,500 people. The sprawling 215 hectare site was first quarried by the Belgian Bertrand family back in 1925 but it is now owned by CBR, part of Heidelberg Cement. Each week they supply Cimescaut's on-site plant with thousands of tonnes of materials for aggregates, while conveyors carry another 30,000 tonnes across the road for shipping to

12.3. With 85 hectares still left to quarry, limestone extraction at Carrières d'Antoing will continue for many years to come.

Ghent where it is used as clinker in the process of making Portland cement.

Up until the turn of the century, the huge stocks of fresh water that accumulated at the bottom of the void were pumped into the river at a rate of 600 to 700 cu m per hour. These days it is not wasted but used for domestic purposes by a Belgian water company. When limestone reserves are exhausted, the quarry will most likely be allowed to fill with water to create a man-made lake for recreational purposes. However, this plan will not be implemented for a long time because with 80 hectares of reserves still to open, CBR still have many more years of extraction left to go.

Caterpillar 994

There are a couple of large, old Liebherr and Demag excavators as well as a newer Terex O&K RH120E, but Belgian quarries tend to favour wheeled power to load their limestone. A few years ago, a diesel-electric LeTourneau L-1000 was the prime mover at Carrières d'Antoing. The machine was capable of shifting 900 to 1,000 tonnes an hour but it was simply not enough to

keep pace with their new and monstrous 400 tonne primary crusher; so CBR bought a 100 tonne Cat 992G that easily lifts 20 tonnes at a time.

When the LeTourneau was finally relieved of its front line duties after 40,000 hours, CBR considered buying a second 992 but, after jetting across the Atlantic to see a machine twice that size in action, they quickly changed their minds. The Caterpillar 994 is not only Cat's biggest wheeled loader but also one of the largest mechanical versions in the world. Both impressed and overwhelmed by the machine's raw power, CBR realised that this single machine could easily do the work of the two current wheeled loaders, thereby, helping to reduce production costs.

Placing the order in 1993 for the very first Cat 994 in Europe turned out to be much easier than trying to get it to the quarry. The main access point was via a narrow, mainline rail tunnel linking Tournai to Mons. Fortunately, there was also a back-door route via a neighbouring quarry, so the fleet of trucks carrying the bits and pieces were diverted next door. When fully assembled, the huge loader was driven to Carrières

12.4. This Cat 994 is Le Barbare, the oldest of two 994s at Carrières d'Antoing. It looks as good as new after 48,000 hours.

12.5. Wheeled loader buckets do not get much bigger than this 19 cu m version.
Lifting in excess of 30 tonnes with ease, it is able to hit production figures of 10,000 tonnes during a single 8-hour shift.

d'Antoing. The massive machine did not leave quarry owners with much change from today's equivalent of €2 million.

Affectionately referred to as Le Barbare (the barbarian), it got its name from a visiting engineer who quite literally uttered this name when he saw it for the first time. The team involved in assembling it were so amused that Europe's first Cat 994 was promptly christened with the name. Its arrival did not go unnoticed and both the dealer and the quarry were inundated with requests to see it. They were even approached by people who had read about it in their local paper and some even tried to sneak into the quarry at the weekend to see the giant machine in the flesh.

Outsiders may view the machine as a bit of a tourist attraction but in the quarry things are deadly serious because the huge loader has a massive job to do. While the 994 can carry buckets with capacities up to 36 cu m, the Belgian machine is fitted with the standard 19 cu m

version, capable of lifting in excess of 30 tonnes at a time. The huge American-made loader lived up to its reputation as a four-pass loader for the quarry's 108 tonne capacity trucks. It easily hit production figures of 10,000 tonnes during a single 8-hour shift.

After 24,000 hours of work, Le Barbare was refurbished in a three-month operation that cost the best part of £700,000. It then went on to notch up a similar number of hours before requiring a major overhaul in 2008, seven years later. Cat dealer Bergerat Monnoyeur tore the monstrous machine into thousands of pieces, but not before finding a suitable replacement for the Belgian quarry. The quarry was aware that their 992G was simply not up to the task of taking its place, so they bought another new 994 – this time a 994F.

The 994 is produced by the same US factory that makes smaller but impressive sisters like the 992 and 133 tonne 993. In 1990, the first of the largest wheeled loaders in the world went into production at

12.6. Dwarfing operator Philippe Coqueriaux, the 994 stands nearly 7 m tall to the top of the exhaust stacks; it is almost 6 m wide and nearly 17 m long with the boom lowered.

12.7. Mechanics from Caterpillar dealer Bergerat Monnoyeur reduce the giant wheeled loader into manageable lumps for transport to the workshop at Overijse near Brussels.

12.8. It took eight dedicated mechanics three months and 5,000 hours to completely rebuild Le Barbare.

Caterpillar's Decatur factory in Aurora, Illinois; then it was known as the 994A. Launched in 1998, the first successor was called the 994D. This model had a single joystick to control both steering and transmission, instead of a steering wheel. The current 195 tonne 994F was previewed in September 2004. Production started the following spring, and since then more than a hundred machines have been delivered.

The total number of 994s worldwide is over 450 and the lion's share of that can be found in the big surface mines in Canada and North and South America. Many of these massive machines work round the clock and notch up between 8,000 and 9,000 hours of operation a year. The number of 994s in Europe can be counted on two hands and the only other 994F models believed to be in Europe work at the Aitik copper mine in Sweden. They are covered in Chapter 13.

As soon as the newcomer was assembled and ready for action, Le Barbare was relieved from its role as prime mover. The changeover may seem like it was an easy task but even when broken into bits the main chassis of

Le Barbare still weighed nearly 80 tonnes. To make matters worse, although it was just 120 km to Bergerat's Overijse base near Brussels, the transport crew were forced to take numerous detours, meaning that the actual route worked out to be nearly 500 km.

Once at the site, eight dedicated mechanics re-worked the pieces of Le Barbare and replaced many others – including the engine – before reassembling the giant jigsaw puzzle. Three months and 5,000 hours of hard graft later, the reconstructed machine was ready for action.

Sylvain van de Caveye regularly spent time behind the controls of Le Barbare before the transformation. He hardly recognised it when it returned. 'It looked and felt like new,' he said. Before the revamp, Sylvain says that he used to get worn out because he needed at least three hands – one for the steering, one to change gear and one to operate the controls – to manoeuvre the beastly machine. 'At the end of a shift I was absolutely shattered,' he says. He reckons the rejuvenated cab is now not only much quieter but also far more comfortable. The addition of the joystick control provides for much easier operation. 'The steering system on both machines is now identical, making it easier to change from Le Barbare to the 994F.'

The Cat Certified Rebuild programme costs roughly fifty-five to seventy-five per cent of the standard US$2.6 million price tag attached to a brand-new 994 wheeled loader. Clearly, CBR saved money by using the programme to revamp Le Barbare but, back at the quarry, Cat's largest wheeled loader was relegated to the role of assistant, behind the newer 994F.

After a successful first fourteen years, quarry director Vincent van Overbeke says that CBR are now looking forward to the next fourteen years. He stressed that their need for two 994s is well-founded. 'We need a production machine and a back-up machine capable of the same outputs,' he states.

Arriving on the back of ten trucks, the new 994F has been given the nickname, Ton-Ton, which is the familiar

12.9. Sylvain van de Caveye spent a lot of time behind the controls of Le Barbare, even when it had a steering wheel. He loves the modifications to the cab and especially the new joystick controls.

12.10. Painstakingly restored to nearly mint condition, Le Barbare is easily capable of taking over when Ton-Ton or the 992G are out of action.

12.11. Ton-Ton is the name of the new 994F. While the total global active park exceeds 450, the numbers of these machines in Europe can be counted on two hands.

French term for uncle. The name was chosen as a mark of respect for the salesman that sold them their first 994 back in 1993. Though that salesman is now retired, the quarry felt that this was a fitting tribute.

Everything about the pair of huge loaders is phenomenal. The new 994F is shod with 3.5 m tall tyres, making the top of the exhaust stack measure nearly 7 m from the ground. It is almost six metres wide and roughly 17 m long with the boom lowered. The bucket of this formidable loader reaches to a dizzy height of 11 m, helping it to easily negotiate the truck's 8 m skip heights.

The power of this machine comes from the same 16-cylinder engine originally fitted to Le Barbare; the difference is that this latest version now features new turbochargers, air cleaner elements and offers an extra ten per cent displacement. The extra features boost the total volume to 78 litres which, in turn, increases the maximum power from 1, 336 hp for the 994A model to 1,600 hp. When stretched, it burns nearly 140 litres of diesel an hour, which makes it just as well that the former 2,960 litre fuel tank has been swapped for a bigger 4,641 litre version.

Steel Protection

Quarry work generally denotes hard work and difficult conditions, and limestone quarrying is no exception. Machines have to combat the effects of limestone splinters, which shred tyres and wear out wheeled loader and excavator buckets. Limestone splinters are so sharp that it is almost possible to shave with them; this means that without steel chains, tyre rubber would be shredded after just a few thousand hours. With a set of four tyres costing upwards of £120,000, chain protection is a must for the machines. Adding 20 tonnes of weight to each machine, this steel protection helps to extend tyre life by up to 15,000 hours. This help comes at a price, however, and that price is almost £120,000 for a set of four chains.

Limestone is not only destructive on the tyres of the prime movers but also on wheeled loader buckets, which is one of the reasons why many quarries tend to have spare buckets on site. When new, Le Barbare carried a spade-type bucket but penetration was not good and the loader struggled to fill it. Operators were also concerned about teeth snapping off and finding their way into the crusher. The solution was that the

12.12. With an operating weight near 200 tonnes, Le Barbare was the very first 994 to come to Europe. Without chain protection on its tyres, the sharp limestone would cut them to shreds in a few thousand hours.

Belgian machine was the very first to get a new bucket with a serrated-edge design; now a popular option on many Cat wheeled loaders today.

The big loaders have huge appetites so three blasts need to be carried out a week – each one releasing nearly 40,000 tonnes – in order to provide enough material for the machines to shift. Although the 994 is easily capable of single-handedly supplying their primary crusher with 1,500 tonnes an hour, there was a problem. The crusher pulverised blocks as large as 1 cu m but some of these larger lumps tended to bridge in the bottom, forcing truck drivers to empty skip contents in stages. This had the effect of seriously slowing down the operation. Using a hook and chain to remove the bridged lumps was not only difficult and potentially dangerous; it also often wasted several hours of valuable loading time. The solution was not cheap. In the end, the quarry purchased a XL1200 hydraulic arm from Sandvik-Rammer for €300,000. Truck contents can now be emptied quickly because the hammer makes quick work of breaking up any blocks that get stuck in the bottom.

Truck Update

With the loading part and crushing part now optimised, quarry management have turned their attention to the ageing fleet of Dresser Haulpak trucks. The quarry has five 445Es. The oldest two came in January and March 1987, two others started work in 1990 and 1991 and the fifth arrived in May 1996. The oldest versions offer a payload of 108 tonnes and they have now notched up a very respectable 60,000 hours.

The Dresser Haulpaks have provided good, reliable service, but along with the rest of their hauling fleet – which includes a pair of Unit Rig MT-2700s from a neighbouring quarry – they will soon need replacing for something more modern. Quarry management have ambitious plans to take production to higher levels. While their primary crusher is capable of handling 3,000 tonnes per hour and a maximum of 6 million tonnes a year, currently they only feed it with 4 million tonnes a year.

The planned increase in production will not only put pressure on their trucks but it will also affect Ton-Ton.

12.13. Some of the 108 tonne capacity Dresser trucks have seen nearly 60,000 hours of active service and they will soon need updating.

Working alone, Ton-Ton currently averages 18,000 to 20,000 tonnes from two shifts per day – one from 5am to 12.45pm and the second from 9.30pm to 5.15am. If they want to take production to higher levels, then at the very least they will need to add a third 8-hour shift, or maintain the two-shift policy and make full use of Le Barbare. This latter option is a real possibility for the quarry and one of the reasons why they gave the older machine a brand-new engine, rather than a revised one.

With a new engine, Le Barbare should easily give them 20,000 to 30,000 hours of trouble-free service.

Extra production will also bring more changes because the secondary crusher only has a capacity of 1,500 tonnes per hour. Material from the primary crusher will need to be stockpiled in an area of the quarry that is currently housing a temporary workshop, which is far too small to accommodate a 994. The quarry's main workshop is on the other side of the rail tunnel; it is just

12.14. The quarry's ageing truck fleet also includes a couple of Unit Rig MT-2700s bought from another big Belgian limestone quarry.

12.15. Cat's 994 is not an everyday sight in Europe. Even rarer still is the sight of two of these machines working together, especially in one of the smallest European countries.

big enough for the 992 to squeeze through. It looks as though a new and larger workshop inside the quarry would be more than welcomed by the quarry's mechanics, especially during the long, generally wet, Belgian winters.

LeTourneau Trio

Just 6 km east of Tournai in the Province of Hainaut is Guarain-Ramecroix – a limestone quarry that knew no equal in Belgium until recently. The site is owned by Belgian Cement Company (CCB), which is a subsidiary of the Italcementi Group. A few years ago, cement and aggregate production averaged 7 to 8 million tonnes a year, and at its peak, roughly one decade ago, production topped 9 million tonnes – at the time a European record.

Even in the final year of operation, it is still one of the largest quarries in Belgium, with recent volumes of 4 million tonnes. The big machines are hard at work deep at the bottom of the 220 m void removing the final 3 million tonnes of limestone before the operation ceases to exist.

12.16. This 992G just manages to squeeze through the main line rail tunnel separating the quarry from the processing plant. The two 994s are confined to the quarry.

12.17. At its peak the Guarain-Ramecroix limestone quarry extracted 9 million tonnes of limestone but it is now in its final year of operation. The LeTourneau wheeled loader – the largest in Europe today – is nothing but a small speck at the bottom of the void.

With limestone supplies fast running out, CCB faced a difficult decision. The company runs a new site 3 miles away at Barry, which contains estimated reserves of 500 million tonnes – enough to keep them busy for the next 60 years. However, the tunnel connecting the two quarries will not be ready for use for at least another decade. After much consideration, it was decided that it would be possible to push ahead and finish the tunnel a bit earlier but at great cost. The only solution seemed to be to form a joint venture with Holcim, the owner of the Milieu quarry on the other side of the main road from Gaurain. Called SCT (Société des Carrières du Tournaisis), the joint venture ensures that the cement and aggregate plants will remain at full capacity until the tunnel and the Barry site are ready.

Connected by an 800 m long tunnel, a conveyor belt relays crushed limestone from Milieu to Gaurain at around 2,500 tonnes per hour. The tunnel will see plenty of action for many years to come because Milieu has another 200 million tonnes of limestone to dig up – enough for the next twenty years.

Limestone quarrying at Gaurain is not the only thing fast coming to an end. The site is home to a fleet of ancient wheeled loaders and trucks, the likes of which we may never see again in Europe, that are becoming fast in need of replacement.

CCB has always worked with some of the largest earthmovers in Belgium and their impressive trio of LeTourneau diesel-electric wheeled loaders are no exception. The trio includes one 125 tonne, 1,200 hp L-1100 and one 152 tonne, 1,050 hp L-1200. Both of these machines are eclipsed, however, by the 186 tonne L-1400. Capable of 1,800 hp, the L-1400 is the biggest LeTourneau wheeled loader in Europe today.

All three models have played a significant role in LeTourneau history. They were originally created by Robert Gilmour LeTourneau, a California-based mechanic who first entered the earthmoving world in 1919 with a pull-type scraper. During subsequent years the entrepreneur built up a thriving business and, in 1923, he launched his first steel-wheeled, self-propelled scraper, followed in 1937 by the first rubber-wheeled version.

After supplying the Allies with earthmoving kit during

12.18. Wheeled loaders are the preferred choice but Gaurain used to operate a 1,200 hp 1998 Demag H 255S weighing 255 tonnes. The excavator now works at the Milieu quarry for SCT (Société des Carrières du Tournaisis) – a joint venture between CCB and Holcim.

12.19. The L-1400 is the largest LeTourneau wheeled loader in Europe today.

World War II, the next big breakthrough for 'Mr Earthmover', as he was commonly known, came on November 17, 1950. The advent of the first diesel-electric drive system completely revolutionised the earthmoving world. Primary power still came from a diesel engine, but the new drive-train featured four new components: an AC/DC generator as the main source of current, DC electric-traction motors in the wheel hubs, electric operator controls and a dynamic braking system.

After selling three of his five factories to Westinghouse in 1953, R G LeTourneau had sufficient funds to perfect the electric-drive concept, and in September 1958, just five years later, he launched the L-70. This machine was a 58 tonne capacity scraper with electric motors fitted into all four wheels.

He then turned his attention to building a wheeled loader with the electric-drive system. After a few unsuccessful designs, LeTourneau stunned the earthmoving world in November 1967 with the launch of the SL-40 – the largest wheeled loader in the world at that time. The massive machine was powered by two Detroit Diesel engines that combined to produce 950 hp and, at the front, the machine wielded a 14.5 cu m bucket.

Eventually, hydraulics replaced mechanical components, and, by June 1967, a project dubbed XL-1 was in full swing. After Mr LeTourneau's death in 1969, the company was sold to Marathon Manufacturing Company in May 1970, operating as the Marathon-LeTourneau Company. The first 725 hp L-700 LeTro Loader started work at the Navajo coal mine in New Mexico.

In 1978, a more powerful 800 hp Marathon-LeTourneau L-800 was introduced to the earthmoving world. At 84 tonnes – a few tonnes heavier than the L-700 – a handful of these machines even made it to Europe. The L-800 model had a good run, the last of a total of two hundred leaving the production line in October 1985.

Long before production of the L-800 ended, the world's largest mines sought more loading power, so the company premiered an even bigger machine in 1978. While similar in appearance to the L-800, the L-1200 was a much bigger machine, weighing in at 152 tonnes. Carrying a 17 cu m bucket, it had a rated payload of 30 tonnes. Offered with a choice of 1,200 hp Cummins or Detroit Diesel engines, the latter of the two was the most popular. Unfortunately, due

12.20. This Belgian L-1200, which may have originally come from the Aitik copper mine in Swedish Lapland, stands forlorn, ready to be collected for breaking.

to the economic downturn in the early 1980s, just a dozen machines were made between 1978 and 1984.

Six of the twelve went to a mining customer in Colombia while the Belgian quarry took the pilot L-1200 and one production machine. Current mine employees have a vague recollection of a third machine that was purchased from the Aitik copper mine in Sweden in 1985. One of the three is believed to have been written off following an accident many years ago. Today, the only machine still on site has long been out of commission and is currently waiting to be collected and then scrapped.

The economy did not officially start to recover until the second half of the 1980s. Once it stabilised, many of the world's big mining operations regained confidence in the market and began to start spending money on new, high-output wheeled loaders. LeTourneau was no exception; the company went back to the drawing board to develop a successor to the discontinued L-1200 and it was not long before the first L-1100 went to work at Newmont Gold in Nevada.

The newcomer was a far more sophisticated beast than its predecessors, featuring clever electronics, on-board diagnostics and an advanced computerised traction system. Even though the L-1100 was lighter and less powerful, it still carried the same size bucket as the L-1200 and it out-performed the older machine with surprising ease.

Today, L-1100s can still be found all over the world and there are even a few that are still active in Europe. Only a few years ago, the Belgian machine from the 1980s regularly clocked up 4,000 to 5,000 hours of operation a year. Now relegated to the role of back-up, the L-1100 may even be allowed to live out the rest of its days in the stockpile area, providing there are no serious problems when the quarry closes.

For a number of years there was no serious rival to this giant loader, but this all changed in October 1990 when Caterpillar launched its mechanically driven 994, putting pressure on Marathon-LeTourneau. Just a year later, the pressure increased further when Dresser launched the 1,350 hp 4000 Haulpak, which, at 150 tonnes, was another mechanically operated monster. LeTourneau reacted quickly and just a month later started testing a

12.21. Powered by an engine capable of 1,800 hp, the L-1400 gets through 150 litres of diesel an hour. The massive 3,000 litre diesel tank is topped up at the end of each shift.

12.22. The massive wheeled loader dwarfs assistant quarry manager Lemmens Regis.

12.23. The L-1400 lifts an average of 40 tonnes at a time and the huge 21 cu m bucket only needs four passes to fill the trucks. Despite being fourteen years old, the largest LeTourneau wheeled loader in Europe is still up to the task of loading 1,500 tonnes an hour.

12.24. The cab is big enough to easily contain a trainer seat but the steering wheel and analogue instruments are obsolete compared with today's wheeled loaders. Even so, operator Carlton Bruno confirms the L-1400 is a machine with phenomenal power.

12.25. Blocks that are too large for the crusher are dropped to the floor from full height and pulverised into manageable pieces.

186 tonne prototype. Fitted with a 21 cu m bucket, it was the largest loader ever to go into production.

This massive machine was called the L-1400 and with it customers had the option to choose between either a 16-cylinder, 1,600 hp power pack from Detroit Diesel or one with the same specifications from Cummins. The outputs of these engines were increased to 1,700 hp in 1994 and then again to 1,800 hp only two years later. Given that the machine can burn through 150 litres of fuel when working flat out, the massive 3,000 litre diesel tank is an absolute necessity.

The first full-production L-1400 was shipped to Australia in January 1992 and the Belgian quarry – which already had an L-1100 and two surviving L-1200s – brought one to Europe in 1997. Despite its ageing technology and fifteen years of hard work, the largest

surviving LeTourneau wheeled loader in Europe is still expected to meet current production targets. With the closure of the quarry looking fairly imminent, management are keeping their fingers crossed that the big L-1400 will continue to perform right up until the end.

Rare Unit Rig Trucks

If the Guarain-Ramecroix quarry is guilty of running their LeTourneau wheeled loader into the ground, then they are also guilty of doing the same with their Unit Rig trucks. These massive trucks are an ageing fleet sourced from Canada and Sweden to help extract the last of the limestone from the Guarain site. They bought four used 154 tonne capacity Mark 36s, capable of 1,600

12.26. The quarry's ageing truck fleet includes some of the rarest Unit Rig Lectra-Haul trucks in Europe.

hp, and an equal number of more powerful MT-3600s, which use the same 1,800 hp engine as the L-1400. These larger trucks carry 172 tonnes. Over twenty-five years old, these are the only working survivors in Europe today, with the exception of a pair of MT-3600s at Titania in Norway.

Until recently, the truck fleet also included five very rare 108 tonne capacity MT-2700s. Originally, CCB lowered the sides of the skips on these trucks to reduce their payload to 91 tonnes, but when they replaced the 1,050 hp V12 engine for a 1,200 hp engine the size of the tipping body was increased back to 108 tonnes.

Now, however, these trucks are no longer needed so two have been sold to the Carrières d'Antoing quarry. The other three have been parked at Guarain for some time. They will probably end up being stripped for parts or cut up for scrap.

The replacements for the axed Unit Rigs are Komatsu trucks – specifically five, 90 tonne capacity Komatsu HD 785s with 1,178 hp that were made in 2009. The L-1400 needs just three careful passes to fill them.

When lifting 40 tonnes at a time, the L-1400's huge 21 cu m bucket fills the larger trucks sufficiently in four passes. Despite its age, the largest LeTourneau wheeled loader in Europe is still capable of handling an impressive amount – around 1,500 tonnes of material an hour.

In addition to the L-1100, the quarry has a more modern prime mover: a Cat 992G. Management are trying to save this machine for the tough work ahead at the Milieu site; there a pack of similar machines load Cat 777 trucks.

Even if the L-1400 does survive until the end of the current project the preference is to scrap the big machine. Once this happens, both the largest LeTourneau wheeled loader in Europe today and the rare batch of Unit Rig trucks that it fills will be relegated to the history books. While all the staff at Guarain will move next door to Milieu, the 992G and the five Komatsu trucks are the only machines that will join them.

LeTourneau L-1800

The L-1400 was not the biggest LeTourneau wheeled loader in Europe for very long because the US manufacturer soon developed a prototype of an even bigger machine, which was to be called the L-1800. The physical dimensions of this wheeled loader were similar to those of the L-1400 but the 23 cu m bucket on the bigger machine could lift 45 to 50 tonnes at a time.

The L-1800 was unveiled in December 1993, just after LeTourneau was bought by Texas-based Rowan Industries. The new wheeled loader king tipped the scales at just under 200 tonnes and was offered with engine options of either 1,800 or 2,000 hp. Fitted with a 35 cu m capacity coal bucket, the first L-1800 was purchased by the Eagle Butte coal mine, located near Gillette, Wyoming.

One of these powerful machines even came to Europe. It was purchased by Boliden for their Los Frailes lead-zinc mine near Seville, Spain. The L-1800 did not work for long after the tailings dam at Los Frailes collapsed in April 1988 and the machine was shipped to the Aitik copper mine in Sweden. Sadly, the Swedish mine never re-built it and eventually the massive L-1800 was sold to a Canadian mine.

Large as the L-1800 was, the engineers at LeTourneau still managed to create an even bigger wheeled loader. Visitors to the 2000 MINExpo in Las Vegas were stunned by the colossal L-2350, which, even today, is the most powerful front-end loader ever put on wheels.

Measuring nearly 20 m long, this incredible machine weighs 265 tonnes, packs a 2,300 hp punch and carries a 40 cu m bucket. The L-2350 is built at a facility in Vicksburg, in the US, on the aptly named LeTourneau road. For now, none of these extraordinary machines work in Europe.

CHAPTER 13

Sweden

Swedish Lapland forms part of the largest expanse of wilderness in Northern Europe.

One of the biggest settlements is the 20,000 strong town of Gällivare, a remote outpost over 1,300 km north of Stockholm. Located roughly 100 km north of the Arctic Circle, Gällivare's temperatures regularly plummet to minus twenty-five degrees Celsius and colder, even with global warming taken into consideration!

The nearby Lapland Airport proudly displays a sign stating that tourists have arrived at the 'Mining Capital of Europe'. This proclamation is further reinforced by a large sign on the road to Gällivare. They are not hollow words on those signs, for Swedish Lapland is home to Malmberget and Kiruna, two of Europe's largest underground iron mines. Both sites are operated by LKAB and the LKAB-sponsored mining museum at Malmberget features examples of how the early pioneers originally used reindeers to pull sledges full of iron ore pellets before the rail lines were built in 1898.

Today's mining operation provides many hundreds of badly needed jobs and there will soon be even more because iron ore is in demand. To mine more iron ore, LKAB need to open up three surface mines just north of Gällivare at Svappavaara. These new mines, called Gruvberget, Leveaniemi and Mertainen, are the first new iron ore mines to be opened in Sweden for more than fifty years.

However, even after all four mines are fully operational they will still be dwarfed by a gaping crater on the other side of Gällivare. Those fortunate travellers who fly over Swedish Lapland into Lapland Airport are afforded a unique glimpse of this gaping crater, which is one of Sweden's best-kept secrets. Called Aitik, a name thought to be derived from the Sami word Attje meaning larder, this hole is Europe's biggest copper mine. One theory is that the area got its name because there was once lots of grass for reindeer grazing.

13.1. With the largest underground iron ore mine and largest surface copper mine in Sweden, Gällivare claims to be the mining capital of Europe.

13.2. A roadside sign for Aitik.

New Boliden is happy to leave the grass to the reindeer because its main focus is the copper content of this particular area of Lapland. During the past forty years the joint efforts of some serious earthmovers have removed over a billion tonnes of rock. The gigantic hole is currently over 3 km long and more than 1 km wide and it is forecast to get even bigger. Ultimately the hole will drop to 580 m by 2025 and perhaps even deeper after that.

13.3. Aitik is the largest surface copper mine in Europe. The old concentrator at the top is about to be demolished to allow the hole to widen.

13.4. Reindeer are a common sight near Aitik. It is thought that the mine got its name from the Sami word Attje, which roughly translates as larder – in this case one with lots of vegetation for grazing.

13.5. One year ago, these holiday cabins were quite literally in the middle of nowhere.
Now, they have been purchased by the mine and will be demolished to make room for further expansion.

While the current void is being enlarged, contractors are opening up a new dig not far from it with a fleet of Volvo EC700 hydraulic excavators. These Volvos would be big machines anywhere else but not here.

The new dig is called Salmijärvi, and at some point in the future it is quite possible that the new pit will merge with the old to create a hole of gigantic proportions. For now, though, the main mine has its work cut out with a series of massive push-backs.

Much has changed at Aitik during recent years. Not long ago it extracted 18 million tonnes of copper ore a year. However, a massive €1 billion investment in machinery has resulted in the volume of copper ore extracted from the pit doubling. When overburden is added to this figure, they need to shift a staggering 90 million tonnes of material a year!

The newly renamed Aitik 36 utilised the bulk of the recent investment to purchase new crushing, internal transport and processing facilities. Thirty-six million tonnes of copper ore is a phenomenal amount and means the crushers need feeding with an average of 100,000 tonnes a day. In order to remove the huge volumes of overburden to expose this vast amount of copper ore, the machines need to dig roughly 170 to 200 m below the surface. Even when they reach it, every tonne of the low-grade ore contains just 32 g of copper concentrate. The truth is staggering: 36 million tonnes of copper ore, once processed, results in only 130,000 tonnes of copper concentrate.

13.6. At 70 tonnes, the EC700 is the largest hydraulic excavator produced by Volvo.

13.7. With an operating weight of nearly 50 tonnes, this machine is one of Volvo's biggest wheeled loaders; at Aitik, however, it is overshadowed by two machines four times its size!

All copper concentrate mined at Aitik is transported 400 km south to Boliden's smelting plant at Rönnskär. The reason for this long journey is that Rönnskär used to be home to a gold mine and although the gold mining operations in that area closed many years ago, Aitik still makes use of the smelting facilities.

Each year Aitik also produces more than 35 tonnes of silver and nearly 3 tonnes of gold, a sufficient amount to pay the salaries of the 450 staff. Aitik is not only a big copper mine, but also Sweden's largest gold mine. Despite laying claim to both of these minerals, the quantities of copper, gold and silver contained in their ore are truly minute.

There is no better place to grasp just how tiny the portions are than at a display at the locally famous Grand Lapland Hotel where a 1.3 kg chunk of copper ore lies in a glass cabinet. Next to this hunk of material is what the mine is able to extract from it – 5 g of

copper (barely enough to fill a teaspoon), a few grains of silver (0.004 g to be precise) and a speck of gold (0.0001 g). The Grand Lapland Hotel even has a model of one of the mine's big Caterpillar dump trucks, which can hold 220 tonnes of ore when fully loaded. When processed, this load can extract roughly 800 kg of copper concentrate, 600 g of silver and 20 g of gold. This sounds like a fair amount until it is explained that 320 tonnes of overburden needs to be shifted just to get at this single truck-load of copper ore! This is a considerable amount by European standards and therefore requires extremely large machines.

Most tourists come to Lapland for the spectacular natural beauty, winter skiing, and the chance to witness the Northern Lights. However, from April until October, the Gällivare tourist office organises hugely popular guided tours to Aitik and every year several thousand visitors, mainly Swedish, sign up for this visit of a lifetime.

13.8. During the past few decades all the copper concentrate mined at Aitik has left the mine by road; however, plans are in place to begin moving the material by rail.

13.9. Aitik's reputation spreads well outside the mine. Every summer, the mine welcomes several thousand visitors who are eager to get close to some of the biggest earthmoving machines in Europe.

This incredible tour begins with a close encounter with the mine's Marion 7400 dragline, which stands as clear proof of Aitik's long-running history of using big machines. The dragline originally belonged to Vattenfall – a big Swedish energy company and its job was to dam up rivers so that power stations could be built. Boliden bought it around 1966–7 to strip off overburden when the mine officially opened in 1968. This machine has attracted attention not only from tourists but also the media. The Marion 7400 dragline has had a film made about it on a national television station and a song written about it by a Swedish singer.

One of the highlights of the Aitik tour is when visitors take a peek into the giant void from a special viewing point at the top of the mine. From

13.10. The Marion 7400 dragline at Aitik originally belonged to a big Swedish energy company called Vattenfall. Boliden bought it around 1966–7 to open up the mine but it does not see much action these days.

here, even Europe's largest earthmovers appear quite small, an impression which quickly changes when the tourists are later brought face-to-face with the massive machines.

Big Bucyrus

For more than three decades Aitik has been a regular Bucyrus customer. Now retired, Torbjörn Kjellsson, who joined Aitik as chief of mine planning in 1984, recalls the very first machine – a Ruston Bucyrus 150, which at the time supported a 'massive' 4.6 cu m dipper! Used for production purposes from 1968 to 1978, and then as a stand-by until 1985, the 240 tonne rope shovel was able to dig 400 tonnes an hour. This early machine was joined in 1972 by a pair of Bucyrus-Erie 280-Bs; four years later in 1976 by a 295-B, and in 1984 by a 290-B1.

Even as recently as 1985, Bucyrus supplied all the

three main prime movers at the mine. Aitik's long-standing relationship with the American company is still evident today. The 295-B, the only Bucyrus model left at the site, is parked not far from the old workshop. Made in 1975, the 295-B was Aitik's biggest machine right up until 1989.

The staff have fond memories of this last remaining warrior from the past. Although it has not seen any active service since 2009, present-day shift supervisor Sture Holmgren reckons it is the best machine that Aitik has ever had. 'It never stopped and I loved operating it,' he says.

Having trundled to its parking spot under its own steam, it can soon be brought to life by a jolt of electricity, but lots of work is needed before it can do any further digging. Although the 16 cu m dipper will not see any more action on the production side, the mine has no plans to get rid of it. When crushed, much of

13.11. Aitik's preference for rope shovels began in 1968 with this Ruston-Bucyrus 150-RB.
Fitted with a 4.6 cu m bucket, this former prime mover was cut up for scrap about five years ago.

13.12. Up until 1989 this 1975 Bucyrus-Erie 295-B was one of the biggest machines at Aitik. The mine owners have yet to decide its fate.

*13.13. With an operating weight of 1,300 tonnes, the 495BII is the largest
Bucyrus electric rope shovel in Europe and Aitik has the only one.*

the overburden at Aitik is ideal for road building and, after some serious refurbishment, the 295-B would be suitable for this job.

Aitik also owns a 1,300 tonne 495BII, which is a totally different machine and a very successful shovel for the South Milwaukee-based manufacturer. With over a hundred out working around the world, not only is it the largest of the prime movers at Aitik, but it is also the only one if its kind in Europe.

Bought new in 2001, when it cost them nearly £8 million, it has a 43 cu m dipper that is so big that an average-sized family car could park inside it. The 495BII can fill the biggest Caterpillar dump trucks in Europe in just two or three passes and in the right conditions it can fill

as many as 100 to 150 trucks during a single 8-hour shift.

All Aitik's massive machines run on electricity, making the current annual requirement 400,000 megawatt hours, or more than enough to power the whole town of Gällivare. Mine superintendent Glenn Nilsson confirms that diesel-powered shovels are easier to move; however, past performance confirms the simplicity, reliability and low running costs of electric motors. It was these factors that tipped the scales in favour of electricity.

A typical hydraulic excavator would have hydraulic rams on the boom, stick and bucket to allow it to flex its muscles and dig. By comparison, the massive boom fitted to the huge electric shovel of the 495BII is static with four chunky steel cables to secure it to the main

13.14. A big machine needs a big dipper. This 43 cu m one is quite possibly the largest in Europe. When full, it holds over 100 tonnes and is capable of filling 218 trucks with 44,690 tonnes of material during a single 8-hour shift.

structure. The mid-mounted handle and dipper are the only parts that move and it is these that do all the hard work. Operation is surprisingly simple, as is grasping the various functions. Pushing the right joystick away from the operator lowers the dipper hoist ropes, while pulling this same joystick back hoists the dipper. The left-hand joystick controls the swing of the machine and also operates the dipper crowd and retract ropes.

Operator Peter Burck reckons that filling the dipper to maximum capacity is not easy; he says that the knack is in finding the right balance of the crowd and hoist ropes for maximum dipper penetration. 'This takes some getting used to,' he says. 'Too much pressure and the machine will grind to a halt, whereas too little will result in a half-full dipper.'

Despite its immense size and weight, the machine judders and jerks as the dipper is teased up through the heap. Fortunately, the suspended platform helps to isolate the operator from the worst jolts. Unfortunately for Aitik,

the machine requires a lot of tender loving care. Cracked track pads are commonplace, and welders are frequently called in to repair chinks in the dipper and boom.

Despite its relatively young age the shovel has suffered a spate of problems. Some of the staff at the mine joke that it should be left at the face when a blast goes off, while others think that despite the problems the machine deserves credit for its unbeatable hard work.

A fully filled dipper can weigh as much as 150 tonnes and will swing at a surprisingly high speed. Operators need to take great care when loading; in the worst case scenario, the dipper can crash into the truck's tipping body headstock, an impact that can easily lead to truck driver neck injury and damage to the 495BII. Also, when emptying the contents of the first dipper the operator has to carefully lower the load as far into the tipping body as possible because when 100 tonnes of material is released into the truck body the cab windows have been known to shatter!

13.15. The dipper on the Bucyrus 495BII wears quickly in the abrasive conditions. This one is off to be repaired.

P&H 4100

The 1,250 tonne 4100 series is P&H Mining Equipment's most successful electric shovel. The company's original production plans estimated building ten new machines a year; however, since its launch in 1990 P&H have built more than 200. The majority of those work in the US, Canada, Australia and South Africa, but they can also be found in South America, China and Russia.

It stands just under 20 m high and 34 m long, and few other European mines can fully utilise a machine that can load 6,000 to 7,000 tonnes per hour. Aitik has the facilities to make the most of these incredible machines and it proudly lays claim to the only three in Europe!

The first machine – a 4100A, was commissioned in 1997 and after 53,000 hours it was given a major overhaul in 2010. Revised in two stages, the electrical upgrade was first. Although the original electric motors were left in place, the drive was changed from analogue to digital. 'Digital drive is much faster and more reliable, and offers more electric

motor control,' says P&H MinePro service engineer Carl Rowley.

After two further months of operation, it was trundled back to the top of the mine for the mechanical overhaul. This included new tracks, rollers, transmission and drive shafts on the undercarriage. After having its dipper, handle and boom removed, the machine was split at the slew ring. The 800 tonne upper structure was then hoisted by a series of massive jacks. A new cab and handle were added before it was put back together again, and finished with a new coat of paint.

Their second P&H 4100 used to work in Spain at Boliden's Los Frailes lead-zinc mine near Seville. The mine hit the headlines in April 1988 following the collapse of the tailings dam when 4.5 million cu m of toxic sludge were released into the River Agrio. Even today it is remembered as Spain's worst ecological disaster. It led to the closure of the mining operation leaving many large machines redundant. Many of

13.16. The 4100 has an operating weight of nearly 1,250 tonnes and is P&H's largest and most popular electric mining shovel. There are nearly 200 P&H 4100s out at work around the globe but Aitik owns the only ones in Europe.

them were stripped and shipped to Aitik, including a huge LeTourneau L-1800 wheeled loader that was never re-built because the parts were sold on to a Canadian mine.

The Spanish P&H 4100 rope shovel was dismantled and transported by boat to northern Sweden in 2002. It was ready for work by 2004. Former Aitik employee Torbjörn Kjellsson remembers it as a logistical nightmare, requiring twenty-five vehicles to haul the super-sized loads over 200 km from the Swedish port to the mine. 'Moving the 80 tonne dipper was the biggest problem,' recalls Torbjörn.

Currently, the P&H 4100 from Spain is undergoing a major overhaul. It has a number of cracks that require attention; some are due to metal fatigue and others have to do with the cold winter weather. Regardless of what caused the cracks, many weeks of continuous welding will be needed before it is ready for action again.

Aitik has earmarked US$9.5 million to update and overhaul both the 4100A and the Spanish 4100. Fortunately, they expect the machines to perform another 40,000 hours each before any further major investment is required.

13.17. This is what the P&H 4100 looked like for the first thirteen years of its life at Aitik.

13.18. Today the P&H has been rebuilt and repainted in this new livery.
The 43 cu m dipper on the 4100A lifts more than 80 tonnes of blasted rock at a time.

13.19. The dipper on this 4100C – which is nearly 5 m long, over 4 m wide and 4 m tall – is easily capable of scooping up a Cat 980 wheeled loader. When full, it holds between 70 and 100 tonnes of material.

13.20. The Spanish 4100 is currently getting a massive multi-million dollar makeover.
The Gällivare fire department is on standby while vital welding work is carried out to repair a major crack near the slew ring.

The third of Aitik's P&H electric rope shovels is a brand-new 4100C – a colossal machine with the price tag to match. The 4100C cost Aitik US$20 million! The bits and pieces for this latest American-made creation arrived in Sweden in October 2009. Using police escort, it was a massive operation to move them all to the mine. The 46.8 cu m, US$1 million dipper was so large that it needed its very own truck. Out of all the supersized chunks of metal, the 78 tonne revolving frame was the biggest single component. The assembly of all the individual parts required the assistance of three big cranes. The project kept twenty mechanics fully occupied throughout the winter.

The building of the 4100C was supervised from start to finish by Aitik's Glenn Hannu, who was responsible for heading the assembly of all the other electric mining shovels. Given the need to get the machine ready for action as quickly as possible, the assembly of the 4100C was particularly stressful. For six months he was on site every day – weekends included. Even when he was not working he was thinking about what needed to be done the next day. 'The components had to be assembled in exactly the right order,' he says. 'Missing a piece would have resulted in major delays to the machine starting work.'

Building a big machine is challenging even in the most forgiving climates, but the Arctic winter succeeded in making an already difficult task even harder. Fierce snow storms and temperatures that dipped as low as minus twenty-five degrees Celsius severely hampered progress. 'Handling a 67 mm diameter hoist rope is difficult at the best of times,' Glenn says. 'But when it is frozen it is like trying to bend an iron bar.' Even fitting the 'smaller' and supposedly simple components, such as the 1.5 tonne track links and securing pins, was much harder than normal.

However, by mid-January 2010, after nearly 6,000 man hours, the 5,500 different parts of the 4100C were all in place. As head of the operation, Glenn Hannu's hard work was rewarded with the enviable task of driving the brand-new machine down into the bowels of the void. After a few hours at the controls it was time to hand over to the instructors for several weeks of tests and training.

It may be highly surprising to learn that this fantastic machine, the largest of P&H's rope shovels, is not powered by a massive diesel engine. Instead, the yellow cable protruding out the rear supplies it with 6,600 volts of electricity – enough to power a small housing estate. Inside the machine room this is transformed into a more

13.21. They try to carry out as many repairs as possible on site.
A spot of welding is needed on one of the massive dippers belonging to the P&H 4100.

13.22. Dippers do not get much larger than this 46.8 cu m version which weighs 50 tonnes even when empty.
Aitik recently bought a spare one for the pricey rate of US$1 million.

13.23. In the vast expanse of the mine it is difficult to appreciate the size of the P&H 4100C, until the operator gets out of the cab to check the exterior.

manageable 600 volts AC, which is then transformed again inside a converter cabinet to 580 volts DC. The new bit on the 4100C is then controlled using a digital drive system. The DC voltage and current is responsible for powering seven electric motors – two each for hoist, swing and machine travel and one for crowd.

While there are several different configurations of the 4100 – 4100C, 4100XPB and 4100XPC – they all use the same basic footprint. The main difference is the 4100C has two swing motors while the XPB and XPC have three for slightly faster and smoother operation. Perhaps the single biggest change when migrating from the A to C series concerns the cab. Described as a Rolls Royce design, from the outside the sun shades make it look like an ice-cream van. On the inside, however, it is plenty big enough for the optional toilet and wash basin. The availability of 120 volt, 220 volt and 380 volt electrical supply means many mines prefer to install other appliances, such as a fridge or microwave. At Aitik all three 4100s carry a mandatory vacuum cleaner because operators are expected to clean the floor at the end of their shift.

There are no foot pedals in front of the operator. Instead, many main functions are controlled from two newly

developed and easy-to-use joysticks. Machine swing, dipper hoist, handle operation and opening the dipper's rear door, are all controlled from these two levers. The six-camera monitoring system is also new. Operator Joakim Eriksson says he can quickly change from one to another. 'They are an incredible improvement,' he says.

Having been at Aitik for fourteen years, Joakim was eighteen when he first got behind the controls of an excavator. Hooked from the first moment he operated the machines he says he has never wanted to do anything else. Joakim's shift involves operating one of two machines – the rope shovel or a wheeled loader – for five hours a day and he absolutely loves it.

Each of Aitik's six big electric shovels is always accompanied by a Cat 980 wheeled loader. Skilfully darting in and out, sometimes between the shovel and truck during loading, the wheeled loader operator's job is to clear away any rocks or obstructions that might damage truck tyres. It also ensures a clean running area for the big electric shovel's tracks to run on and often assists in moving the power cable.

The two machines always have a team of three operators –

one person for the shovel, another for the wheeled loader and a third on break. Shovel operators spend no more than two hours at a time at the controls. This policy was instituted following a spate of back and neck problems with rope shovel operators. They found that this could be prevented when drivers spent just two hours at a time behind the controls. After their break, they then spend the next two hours in the wheeled loader.

Arctic winter nights are long. However, Joakim says it makes no difference whether he works during the summer or winter. 'Snow makes the job a bit harder, especially at night. But this machine has good lights, so if you like your job then it makes no difference whether you work during the day or at night.'

The dipper on the 4100C is capable of scooping up 70 to 80 tonnes of material as it rises through the heap of blasted material. The Aitik mine owners declined to specify a new P&H option to optimise the digging process. 'We think operators should be in control of how much can be put in a dipper – not computers.'

In a good bench, the 4100C can load twenty-five trucks an hour – sometimes thirty; at the absolute best, they can load around 6,000 tonnes per hour but this requires plenty of well-blasted material and a plentiful supply of trucks. The burning question now is we will ever see machines of this size anywhere else in Europe? 'Unlikely,' says Carl. 'Very few European mines will warrant such a machine.'

Blue Diamonds

For many years, Demag's 540 tonne H 485 was the world's largest hydraulic excavator. The prototype used to work for Coal Contractors in the UK and when it was shipped off to Canada, Aitik became the owners of the only two H 485s in Europe. The Swedish mine owned a pair of electro-hydraulic front shovels. The first was commissioned from Mannesmann Demag in October 1989. With a cost of around £2.8 million, it was affectionately referred to by mine staff as the Blue Diamond. The owner was clearly impressed because a second H 485S was added just two years later.

During the brief time these two machines were

13.24. The H 485 was once the largest hydraulic excavator in the world and Aitik had the only two in Europe.

13.25. The newer of the two machines was destroyed by a landslide. It was slowly stripped of its vital organs to keep the survivor operational.

operational they handled seventy per cent of the 30 million tonnes of overburden moved each year. Unfortunately the life of the second machine was abruptly cut short following a massive landslide in 2004. Miraculously the driver escaped unharmed, but the excavator was severely damaged where the landslide plunged into its front. The repair costs were so high that the mine officials were forced to scrap it, but many of its parts were salvaged to keep the older machine running. The most obvious example of this could be seen when examining the rear counterweight of the survivor as the number on that particular part gave away the fact that a swap had occurred.

Seventeen years of round-the-clock operation and more than 73,000 hours eventually took its toll on the large Demag H 485 excavator. Machines of this size creak and groan at the best of times but this machine was showing serious signs of failing. Operator Kent Isaksson confirms good and bad spells. On a good day it could still handle some fairly impressive volumes but on a bad one it was out of action for many hours.

Service contracts are generally quite expensive. The ten-year contract with Demag was terminated in 1999 because mine management felt it was cheaper to do the repair work themselves. The Demag H 485 needed a lot of work. The hydraulic system was the single biggest weak point and repeated hose failure needed serious attention. A hydraulic system on any excavator contains many thousands of litres of oil – 8,000 litres with this one. The system on this Demag H 485 was clever enough to be able to detect a major leak which caused it to shut down automatically rather than bleed to death. However, the state of the upper structure and engine room clearly confirmed the source of the problem.

During extremely cold conditions it was a little slower on cycle times, so they tried to run the Demag non-stop to maintain the oil at the optimum temperature. Unfortunately, repeated hydraulic hose failures led to a lot of downtime. 'It cost us a fortune to keep it running but I still have fond memories of the machine,' says Kent.

13.26. On a good day the excavator was able to load nearly 2,500 tonnes an hour and it was easily powerful enough to prise partly frozen glacial drift out of the bench without blasting.

13.27. Although the machine was dwarfed by larger electric rope shovels, the 26 cu m shovel bucket attached to the Demag H 485 still measures over 5 m wide.

Old as it was, the Blue Diamond really did play a vital role at Aitik. Ideally suited to peeling off the 20 m deep layer of glacial sediment at the top of the mine, the Demag excavator was so powerful that it could often manage this directly. However, during cold winter spells, it was not unusual to need a blast to loosen the frozen material. The H 485 was a fantastic machine for handling large rocks and loosening material halfway up a bench. You simply cannot do this with an electric rope shovel.

After it spent a year parked at the top of the pit, Aitik management finally decided to scrap the old Demag H 485 excavator. The ageing giant has since gone on to become one of the world's largest hydraulic excavators – the Komatsu PC8000 Super Shovel.

Komatsu PC5500

Aitik also has two other large excavators that are products of the evolution of the 1995 design of the Mannesmann Demag H 455S, which was a 500 tonne machine fitted with a 25 cu m bucket. Aitik was able to test one of the first machines in a trial run from September 1995 to November 1997.

When Komatsu took over in 1999, the company did not just give the H 455S more power and increase the operating weight, they also gave the excavator a 26 cu m bucket and a new name – the PC5500. The bucket size then increased to 28 cu m in 2006.

Aitik has two of this newest design. Now called the

13.28. The PC5500 is the largest Komatsu excavator in Europe. The Aitik copper mine has two of them.

13.29. With an operating weight of more than 500 tonnes, the PC5500 dwarfs the Cat 980 wheeled loader.

Dash 6, it was created at a factory in Dortmund, Germany. The Dash 6 is thought to be the largest Komatsu hydraulic excavator in Europe today, weighing in at 520 tonnes.

The PC5500-6 model comes with two options for power supply: they can either be supplied with two 1,260 hp, 12-cylinder Komatsu diesel engines or they can run on electricity. The diesel option produces maximum power at 1,800 rpm and the two 38 litre engines are fed from a 10,800 litre fuel tank. Swedish companies usually prefer electric machines whenever possible, but in the case of the Dash 6, the owners of Aitik had originally planned to buy the diesel option. The reason for this decision is that electric machines are slower to move because they require a wheeled loader to help shift the cables that supply the electricity. However, the preference changed following a trip to South Africa in which the company discovered the downside of

diesel. 'It takes nearly an hour to fill a 10,800 litre diesel tank,' says mine superintendent Glenn Nilsson. In the end, Aitik decided to stick with having an electric motor to power their Dash 6.

Inside the Komatsu excavator are two Squirrel cage electric induction motors, each able to generate 900 kilowatts – the equivalent of 1,206 hp – which is only slightly less than their diesel equivalents. The major drawback with electricity as a power supply is the extensive network of cables. Although mines were once able to transform 10,000 volts down to 3,500 volts for the old Bucyrus-Erie 295-B, all machines currently take 6,600 volts. One major side-effect of this is that Aitik is riddled with an extensive network of cables and sub-stations.

Moving any machine a great distance is a frustrating and time-consuming task. These electric machines require at least one, sometimes two, wheeled loaders to

13.30. Electricity powers all six of the prime movers at Aitik, meaning that moving them – and their cables – takes careful planning.

help to move the cables. Unfortunately, the cables easily snag, severing the connection to the excavator and resulting in a motionless machine. Another major problem with electric motors is that the wait time for an electrician can be a long ordeal. One simple solution that might relieve both problems would be to install a portable diesel-powered generator behind the excavator – something that Aitik is currently considering.

The operator of a PC5500 has a long climb up the stairway of the excavator to the operating platform. Once there, however, they want for nothing. At the rear of the spacious cab is a small kitchen, complete with wash basin and fresh-water tank. All controls are neatly packaged on the left of the operator. The only clutter to the right of the operator is a safety lever to engage or disengage the hydraulic functions.

The front shovel construction comprises eight hydraulic cylinders – two for the boom, two for the dipper, two for the bucket and two for the shovel bucket open/close. Most functions are controlled from two joysticks but the PC5500 also has four foot pedals. These are used to operate the two shovel bucket cylinders and change the travel direction of the tracks.

Even though the two excavators are identical, operator Leif Palo prefers the newer one.

'The controls shouldn't feel different but they do,' he says. 'The Dash 6 seems a bit slower, but much smoother.'

However, he is not too keen on the climate control system in either machine. 'The fan in the cab is very noisy and it is really difficult to find the optimum temperature. It is always either too hot or too cold with no middle ground.'

During his nineteen years at Aitik, Leif has operated all machines but the two PC5500s are his favourites. 'They are much more nimble and fun to drive. I can

13.31. The bucket measures over 4.5 m wide and weighs 24 tonnes even when empty. It easily lifts 50 tonnes when full and, in good conditions, can sometimes lift as much as 70 to 80 tonnes.

13.32. Operator Leif Palo thinks that it is much harder to operate the PC5500 than the mine's electric rope shovels.

move quickly to the left or right, whereas it takes much longer to change position with a rope shovel.' One other feature that Leif is impressed with is the ability to use the bucket to assist with turning the machine. 'This is something else you cannot do with a rope shovel.'

Leif also likes the fact he can use the shovel bucket to pick up large rocks that fall into the truck loading area. 'It's so convenient to be able to do this myself without having to call in a wheeled loader.'

However, he concedes rope shovels are much easier to operate. 'In order to operate the PC5500s, I have to make many simultaneous joystick movements. Rope shovel operation is so much easier because all you have to do is power the dipper up through the material.'

Leif used to spend many hours on the Demag H 485 before it was scrapped. Although the PC5500 cab is a huge improvement, there is one thing he misses about operating the Demag. 'It had a longer reach.' He says. 'A couple of metres really make all the difference.'

The two Komatsu excavators are weathering the rigours of their intensive jobs well. On average they see 5,000 hours of active duty each year. The mine is keen to test their limits and see whether they can make it to their predicted lifespan of 70,000 hours.

Big Cats in Europe

Aitik's machine fleet also includes a number of Caterpillar excavator and truck models. The mine owns two Cat 994s that are two of the world's largest wheeled loaders, weighing 195 tonnes each. This powerful machine first went into production in 1990 and since then it has proved a popular choice with many of the big mines. There are nearly 500 working around the globe but only a dozen of them are owned by European mines.

One of the big features of the Cat 994 is that it is flexible. Unlike Aitik's big electric rope shovels, the 994 can travel easily and quickly from one job to another. Whether the job is cleaning the ramps after a blast or extracting copper ore and overburden that is out of reach of the shovels, this machine can handle it. Powered by a 1,577 hp engine, the newer 994F has no problem helping out as a primary loader when one of the shovels is out of action. The 22 cu m bucket can lift 35 to 40 tonnes of material at a time, easily filling one of the biggest trucks in only six passes.

13.33. With an operating weight of 195 tonnes, the 994 is one of the largest mechanical wheeled loaders in the world.

13.34. The 22 cu m bucket fitted to this 994F is easily capable of loading up to 35 to 40 tonnes of material at a time.

13.35. When working flat out the 994F can burn nearly 140 litres of diesel an hour, meaning that the 3,800 litre diesel tank needs to be topped up at the end of every second shift.

13.36. The tyres stand over 3.6 m tall – even bigger than the ones fitted to the Cat trucks. Chains are attached to provide extra grip and protect tyres from damage.

Each chain on the 994F weighs 5 tonnes, which helps boost the operating weight of the machine to 215 tonnes. These chains provide extra grip for difficult terrain and help protect the expensive tyres from unnecessary damage. While it might be slower than the bigger electric shovels, the 994 is still a fast loader. The wheeled loader is also a thirsty machine capable of consuming upwards of 140 litres of diesel an hour, which easily drains the fuel tank after two 8-hour shifts.

One safety feature that is included on this machine, as with all the big trucks, is the automatic fire extinguisher for the engine compartment. If this feature had not been standard, then Aitik might have requested it for the simple reason that if this machine goes up in smoke, then it will cost a cool US$3 million to replace it!

There is also an older model 994A that has been on site since September 2000. Sadly, after nearly 45,000 hours of operation, its prime-moving days are nearly

13.37. This big Cat mechanical wheeled loader spends most of its time on road-building duties. However, it is powerful enough to join the loading team when a big shovel is out of action.

13.38. This older 994A is a frequent visitor to the mine's workshop. There is a good chance that it will be scrapped and replaced by a brand-new 994F.

over. Aitik's mine management are replacing it with a brand-new Cat 994F.

Aitik is also the domain of the largest Caterpillar trucks in Europe. Their 793s are the only ones in Europe and the mine does the best it can to represent these mammoth machines. As visitors turn up the approach road to Aitik, they are immediately confronted with a huge true-to-scale painting of one of them, which was made to celebrate the jubilee edition of a cross-country ski race held on the first

Saturday of April at Dundret, the mountain that rises above Gällivare. LKAB and Boliden sponsor many events and clubs in the area and this particular painting by a local artist was proudly displayed along part of the race course. After the event, it was decided that the painting would be moved to the entrance road to the mine.

The somewhat unusual artwork certainly highlights the size of the 793, but there is no substitute for seeing one in the flesh. Made at the Caterpillar factory in

13.39. This huge painting was originally created for a winter ski race, now it takes a prominent place on the entrance road to Aitik.

13.40. This fleet of twenty-five Cat 793 rigid haulers are the only ones in Europe. They can carry 220 tonnes of material when full.

Illinois, the machine is truly enormous. When full with 220 tonnes of material, it weighs nearly 400 tonnes.

Aitik currently own twenty-five Cat trucks in their fleet, including five 793Bs, six 793Cs, eight 793Ds and six new-style 793Ds that had to be leased. All these trucks cost the best part of US$2.5 each and all are fitted with the optional light package because they are expected to work 24/7, even during the long and dark Arctic winters.

The 78 litre engine produces a peak output of over 2,400 hp and burns 400 litres of diesel an hour when climbing out of the mine with over 200 tonnes of material. Everything about this truck is immense, with the exception of the cab which provides just enough space for the operator and a small trainer seat.

13.41. The 793 is a massive truck and Aitik has the only ones in Europe.

13.42. Aitik's fleet of 793s includes eight of these new-style 793Ds. The sleeker exterior styling gives it a more modern look.

13.43. At the start of each shift, drivers have to inspect the tyres, check for oil leaks and assess the general condition of their trucks. The tyres on these trucks weigh 4.5 tonnes and stand 3.5 m tall.

13.44. Fully loaded, the 793 can burn up to 400 litres of diesel an hour when travelling up the haul roads.

Each spring, the mine recruits seventy to one hundred temporary staff to take over while the full-timers take a summer holiday. The month-long training programme used to comprise a few days of theory and over a week of practical in a play area at the top of the mine. Then after just three weeks accompanied by a trainer, trainees would be ready to operate the machines alone and join the rest of the haul team.

However, the drawback to the first stage of the training programme was that it used to tie up trucks, a luxury that Aitik could not afford. In order to solve this problem, the mine owners recently installed a truck simulator. There is good reason why only ten people in the Aitik staff hold the key to the ultimate computer game – the technology cost the mine nearly £400,000! Reports are that the simulator has proven to be worth every penny: last spring they successfully used it to train nearly a hundred recruits.

The creator of the simulator was an Australian-based company called Immense Technologies. Their technicians spent nearly six months filming all the machines at the mine, as well as haul roads and waste dumps. Those images were then used to make the 3D images.

Trainees need to be concentrating fully when utilising the simulator because they are bombarded with masses of information. Sitting behind the controls is no different to being in the operator's seat of a real truck: the technology is incredibly realistic. It even includes an operating platform and radio microphone. One trainee actually suffered from motion sickness while engaging in a simulation because of how realistically the jolts and bumps had been reproduced. Instructors even have the option to select the weather patterns as well as whether they want the trainee to simulate day or night shifts.

The huge investment is clearly paying dividends with fewer trucks needing to be used for training purposes. The most difficult part of the training is getting drivers to trust

13.45. The new truck workshop is huge, easily capable of holding all twenty-five of the Cat 793Ds. It can also take the larger Caterpillar trucks.

the mirrors and get used to the width of the vehicle. The instructors seem to have reached the consensus that women are much easier to train than men. 'Women listen to what we have to say,' says Magnus Keskitalo, one of five instructors.

The simulator is not only of use to novice truck drivers. Seasoned operators are able to swap the seat and controls of their normal machine for those of a P&H 4100 or a Cat 994 wheeled loader or a Bucyrus 495. However, adding a new machine to the system is very expensive: it took the Australian company one and a half years to make a programme for the P&H 4100 excavator. Technicians estimate that it takes a year to make a new truck programme.

However, even though the real trucks are adorned with huge mirrors, there are still quite a few blind spots. This is one of the reasons why flags carrying masts rise well above the mine's lighter vehicles. Truck drivers spot this before they see the vehicle. In an attempt to tackle this visibility issue, the mine opted to put cameras on their newest 793Ds. The

drivers thought they were great when they first arrived, especially the reversing camera.

Unfortunately, an unforeseen drawback to the cameras is that they quickly become dirty and need regular cleaning.

13.46. All novice truck drivers spend time behind the controls of Aitik's simulator. This ultimate computer game is very realistic.

Also, images from the cameras are displayed on a single in-cab monitor causing drivers to have to switch between different camera shots; this is difficult in practical situations because it is distracting to the operator. Most admit that they would probably rely more on the cameras if provided with three monitors but space is already at a premium and there simply is no more room.

Wonder Women

Despite the immense size of the machinery in the fleet at Aitik, it is not macho men, but women who sit in the majority of the trucks. In fact, over thirty per cent of the workforce at Aitik is female and the mine is keen for more to join. Mine officials would like to gradually increase the number of women drivers to fifty per cent if possible.

Shift supervisors openly admit that female operators not only take more care of the machines they operate but that they also take less risks: two traits which are preferable when considering the price tag on the heavy machinery at the mine. Women also help to moderate the language that courses through the two-way radios during operational hours!

For the past five years, thirty-year-old Johanna Storm has sat behind the wheel of the largest Caterpillar dump truck in Europe – a far cry from the waitressing job that she held in a Gällivare restaurant. 'I was looking for a challenge,' she says, 'something completely different. This beats waitressing and is much better paid.'

She also gets more holidays and certainly more breaks to help her relax after driving six hours and forty-five minutes of each eight-hour shift. Johanna reckons the massive truck is even easier to drive than her car. It is true that sitting behind the wheel it even feels like a car – albeit a big one – and this is because Cat altered the dashboard layout while developing the latest version of the 793D.

The mine's truck drivers work in five-week cycles. A week on the early 6am to 2pm shift is followed by seven days from 2pm to 10pm. After breaking for a week, the operators tackle a week of nights from 10pm to 6am before beginning the whole cycle again with the morning shift. All drivers seem to have a preference for a particular shift. 'I hate early mornings and prefer to work at night. It is great during the summer when the sun does not set for a couple of months,' Johanna says.

Johanna has plenty of experience with both older- and newer-type 793s but clearly prefers the newer, much quieter

13.47. Anna Hagström is one of many women who have been trained to drive the big Cat trucks. 'It is fabulous to drive,' she says. 'Just like a really big but nimble go-cart.'

ones. The new trucks are so much more powerful, something she has time to notice on the long haul to the waste dump.

Many people might not consider Johanna's job to be particularly stressful. Her main task is to complete the journey to and from the waste dump as quickly as possible, ensuring that the cab and mirrors are clean at the end of every shift. On the surface, having such a simplistic job description might make it appear that her job is also unexciting. Johanna goes so far as to admit, 'Driving a truck certainly gives you a lot of time to think.' She passes time by listening to lots of music and keenly following the two-way radio conversations throughout the rest of the mine. 'Work colleagues are really important.' She says. 'We are like one big family here at Aitik so there is always something to talk about.'

Johanna reckons the hardest part of her job is in the winter when the haul roads are covered in ice and snow. 'Trucks can slide around a bit in the winter and we have to be very careful.'

For the past eight years, former sweet shop assistant and retirement community worker Anna Hagström has sat behind the wheel of one of Aitik's big trucks and she loves it. 'It is fabulous to drive,' she says. 'Just like a really big but nimble go-cart.'

While there is a lot less traffic around Aitik than on a public highway, Anna's preference is for the night shift when the trucks are the only vehicles on the haul roads. She also relishes the dark and bitterly cold winter mornings. 'By the time I have checked the outside of the machine and climbed into the cab I am freezing.'

With the heater turned on full blast and the radio for company, things quickly improve and she is really in her element when the day begins to dawn. Although the sun does not make it above the horizon for several weeks at the turn of the year, there is still some light. 'The scenery is stunningly beautiful at this time of the day and there is no-one else around.'

Originally born in Gällivare, two years ago Anna moved further south to Luleå to study. She currently lives with her grandmother back in Gällivare where she has plenty of free time to pursue her art education outside her unique job. 'When I am trucking there is plenty of time to plan my art programme,' she says.

The majority of Aitik's women sit in the trucks but half a dozen are also trained to operate the big rope shovels. Gerd Martinsson is one of them. After joining the mine as a truck driver in 1995, it took just four

years before she was operating the P&H 4100s. 'I soon got bored of driving the trucks,' she says. 'Time passes more slowly when driving trucks than on the shovels.'

Today, she feels completely equal to the male operators at Aitik. In fact, Gerd is able to operate all the machines at the mine except one – the 200 tonne Cat 994 wheeled loader. 'They want me to learn but I am not sure,' she says. 'It is so big and powerful.' Despite her protestations, there is no doubt in the minds of many that no machine is unconquerable for women like Gerd.

Huge Unit Rig Trucks

While the Cat 793 is definitely one of the biggest haulers at Aitik the mine also owns quite a few Unit Rig-made Lectra Haul MT-4000s – the only models of this kind to be found anywhere in Europe.

Sixteen of them used to work at the Los Frailes lead-zinc mine owned by Boliden and located in Spain. Unfortunately for the Spanish, the collapse of a tailings dam heralded the death of that mining operation. After a year of cleaning up the mess, all sixteen trucks became redundant and were eventually transferred to Sweden.

Aitik had plans to reassemble the first big truck in June 2002 and the rest soon after, but Spanish bureaucracy made this impossible. As a consequence, the first batch of four trucks was still parked in Seville harbour at the end of May. After the initial stalemate, progress was made and by September the last of the three shipments had left Seville for the Swedish port of Luleå. The boat ride turned out to be the easy bit, while the hardest leg of the journey was the 200 km road trip to the mine.

Swedish police were needed to escort the vehicles carrying the parts of the trucks to the mine: even then it took over seven hours to transport each load! The tipping bodies posed the biggest transport nightmare. Most large tipping bodies would be supplied in two halves and then welded together on site. This was not an option with the Lectra Haul bodies and so 8 m wide parts proved a nuisance to transport. It is not surprising then that Aitik's mechanics were disheartened by the terrible state of the trucks. Fortunately, by the time they had finished, the Unit Rigs looked like new. The job was completed by changing the livery of the machines from white to yellow.

In December 2004, one of the eight trucks perished in a ball of fire. The operator, in his haste to get out of the

13.48. Of the sixteen Unit Rig Lectra Haul MT-4000s moved from Spain to Aitik, only eight were ever rebuilt. This is one of just three survivors.

cab, not only forgot his boots but also forgot to switch off the engine. Burning fuel continued to feed the fire, which quickly engulfed the rest of the truck. The tyres smouldered for another two weeks. Since then all Aitik vehicles have been mandatorily fitted with automatic fire extinguishers.

Only three of the Unit Rigs remain functional and they are on standby in case some of the Cat 793s are out of action. The huge graveyard of parts scattered behind the old workshop has all been cleared now but high on a bench on the other side are the remains of one of the huge radiators belonging to a deceased Unit Rig. Proudly wearing the numbers XVI, fifteen years ago the truck to which it was plucked from was driven by Carl XVI Gustaf, King of Sweden. Today, this is the only vital organ of this truck that remains.

Although Unit Rig stopped making the trucks in the early 1990s, the MT-4000 was offered with four engine options: a 2,000 hp Detroit Diesel engine, a 2,500 hp Detroit Diesel engine, a 2,000 hp Cummins engine or a 16-cylinder 6 tonne 2,467 hp MTU.

Incredibly, after more than 35,000 hours of work all three surviving MT-4000s have each covered more than 400,000 km! While one uses the latter engine option, the other two have Cat-powered stickers stuck to their bodies because they had their former engines swapped for 2,415 hp 78 litre Cat engines.

During the winter, all parked trucks are connected to an electricity supply for the engine and cab heating systems. With winter temperatures regularly falling well below freezing, this is no luxury measure. Without electricity keeping the engine warm at night, it would be almost impossible to start any of their trucks on really cold winter days.

The MT-4000 is a diesel-electric truck. The diesel engine powers a generator which provides electricity to DC motors in each of the rear wheels. Unlike the Cat 793 which offers full hydraulic braking at the front and rear, the electric braking on the Unit Rig is only available at the rear. The mechanically operated front disc brakes must only be used at low speed around the loading shovels. If these 1.17 m diameter single discs are used at speed then it is quite possible that they could catch fire because there is no cooling system. To reduce the risk of damage the mine has limited the top speed of the three survivors to 32 km/hr. An electric retarder prevents it from going any faster.

The Unit Rigs are not popular in the winter when the haul roads at Aitik are covered with snow and ice. Most operators have an understandable preference for the all-wheel braking found on the Cat trucks and they complain that the brakes on the Unit Rig are inadequate. Experienced operators are needed to work with the MT-4000 during the winter: operators must know that the front brakes should only be used as an emergency. If drivers hit them when going downhill, then the discs will need checking at the workshop.

Water Babe

Dust is not a problem during the long, dark and fiercely cold winters that characterise Swedish Lapland. During the brief summer spell, however, Aitik's haul roads soon dry up and strong winds blowing down from the north propel that dust into the air.

The mine used to rely on a local contractor to keep the level of dust under control. Unfortunately, a tractor and farm slurry tanker combination, with a working width of just 2.5 m, was not very effective on their 25 m wide haul roads. On a bad day, rising dust clouds could be seen from the town of Gällivare. It became obvious that something else was needed to keep on top of dust, especially as the mine grows deeper and wider and the haul roads become longer.

Aitik shift supervisor Börje Hansson was given the task of searching for something more effective. He was immediately keen to learn more about the American-made conversion kits for big rigid dump trucks. The company he was specifically interested in, called Mega Corp, was based in Albuquerque, New Mexico. Their product was dubbed Mega Truck Tanks (MTT) and Börje just had to jet across the Atlantic to see their modified Caterpillar trucks in action. Impressed by what he saw, he started looking at the truck fleet at Aitik to see whether they could sacrifice one of their older models.

All the Cat 793s at Aitik were desperately needed for hauling duties so he had to look elsewhere. He soon found just what he was looking for – a 1989 Cat 789B with 56,000 hours on the clock. After removing the tipping body, Mega Corp sent a mechanic to supervise the rest of the work,

13.49. This is the largest water truck in Europe, holding 132,000 litres of water. When all four nozzles are open, it empties in just thirty minutes.

13.50. It takes just ten minutes to refill the tank from a special filling station.

including the attachment of the new Finnish-made water tank that had to be bolted to the rear of the truck. Power to the hydraulically driven water pump comes from the oil normally used to raise and lower the tipping body rams. At maximum flow it is capable of delivering 5.6 cu m of water an hour to the dusty roads of the mine.

Several weeks and €300,000 of modifications later, the modified Cat was ready for action. The result has no equal in Europe. It is a truly formidable sight: a water truck capable of covering the massive Aitik haul roads in a single pass. During peak periods, the 132,000 litre water truck works two shifts a day – from 6am to 3pm and from 3pm until 12pm. Nights are generally quite damp so it is not needed.

This lack of a night shift is one of the reasons why driver Camilla Enback came back to the mine. Though she may be new to the water job, the construction and operation of the truck holds no secrets. Nine years ago, she spent many hours in the driving seat of this particular model during its rock-hauling days. While she enjoyed the driving, Camilla could not stand the night shifts and, in the end, she left her job at Aitik because of them. However, the mine would seem to have magnetic powers because she came back a year later, having taken a job on a drilling rig. Unfortunately, the night shifts got the better of her and six months later, she threw in the towel once again. 'I love trucks and drilling,' she says. 'Even the early mornings are not a problem. I just hate night shifts.'

Camilla joined her husband's truck business for a few years but when the economic downturn came she once again approached Aitik to see if there were any day jobs. She was in luck, with the advent of the new water truck the mine was in need of another driver.

The cab is still noisy – meaning she normally has to wear ear plugs – and the layout has remained largely unchanged. There is, however, one exciting exception – an electronic control box has been inserted on the right-hand side. The control box activates the water pump, while the rear nozzles are engaged through four rocker switches: two in the centre and two on the sides. Engaging the middle two switches usually produces enough water to soak the large Aitik haul roads but some of the exceptionally wide roads require all four. 'The tank is empty in about thirty minutes when all four are engaged,' she says.

The process of refilling the water tank takes ten minutes. A filling station had to be specifically engineered for this purpose. All Camilla has to do is align the truck through a specially made gateway until the gaping fill hole at the top of the tank is under the water supply pipe.

Camilla clearly loves her job. She and her unique vehicle receive an amazing amount of attention. 'Everyone loves the water truck and they are always pleased to see me,' she says. 'Sometimes a shift supervisor or rope shovel operator will ask me to go to a specific area, but most of the time I decide which road to do. There is no stress and more importantly no one breathing down my neck.' Travelling at 15 km/hr, sometimes faster when going downhill, she can comfortably cover all the haul roads on site with three tanks in just three hours.

During her shift she aims to do about six tank loads – including breaks. Camilla takes great pride in her job and she tries to be courteous to other drivers by remembering to switch off the pump when passing another truck or service vehicle. She admits that there have been times where she has forgotten. 'I did this once when I passed an excavator on repair. The mechanics were not too pleased!'

While the water truck has seriously helped with the dust problem at Aitik even Europe's largest water truck is not enough on the seriously hot and dry days. The dust has to be controlled with a mixture of salt spreading and water from the water truck. By spreading salt over the roads first, a crust is created which helps to keep the dust down for longer. Camilla jokes that she does not appreciate too much dry weather. 'Bring on the rain! If it rains, then I do not need to work so hard!'

However, even on wet days she is not able to sit back and relax. If the water truck is not needed, then there is always something else that will keep her busy. During the long Arctic winter she plans to learn to drive a wheeled loader because the mine will not need the water truck.

Camilla thinks this trade-off will suit her well. 'The wheeled loaders are part of the road gangs. They only work a daytime shift so I can hopefully avoid night shifts all year round. '

The act of spreading water all day, every day is punishing on the truck. So much so that it will need replacing in another three or four years. 'All our remaining 789s will be too old for this work so it will probably be the bigger Cat 793 that will be modified next,' says Camilla.

In the meantime, the Cat 789B mining truck is ideal for the 24 m wide haul roads used by the mine's Cat 793D trucks. Driving slowly and manipulating the side nozzles will boost the working width to an incredible 32 m. If the larger Cat 793 truck is the next available choice for modification, the span of the water nozzles will be easily wide enough for the new fleet of even bigger Cat trucks which have just started work at Aitik.

The Future

Originally, US$1 billion had been earmarked for further expansion and, so far, not all of it has been spent. As it stands, there are still a few more machines that Aitik would like to acquire in the future.

13.51. Many miles of conveyor belts have been installed to speed the transport of copper ore to the new processing facilities and relieve the pressure on the truck haul team.

13.52. Aitik's massive new workshop provides a warm, dry atmosphere for mechanics to service and repair many of the mine's big trucks and machines.

Currently, they have twenty-five Cat 793 trucks, which may sound a lot but the reality is that it takes more than that to keep pace when all six shovels are working – especially considering the long hauling distances. Several miles of conveyor belt have been installed to help take the pressure off the copper ore team but colossal volumes of overburden will still need to be moved. The capacity of the largest Caterpillar truck in Europe today is no longer sufficient for the amount of overburden that needs to be shifted, so it is a good thing that Aitik has just bought a number of even bigger Cat 795F ACs.

For the moment, they will have to make do with their new-style 793Ds, which are leased until the ten 795F ACs are assembled. Aitik is the first mine to own these massive Caterpillar trucks and because they are the first, Caterpillar will have to provide a lot of support to the mine in case there are any last-minute teething problems. Almost 2 m wider than the 793, the first two have just started work, while the other eight should be fully operational from 2012. Mine superintendent Glenn Nilsson reckons the 795F will be the perfect match for their four big shovels. 'They need five passes to fill it with 315 tonnes,' he says.

However excited the mine officials might be about their new fleet of trucks, there are a number of logistical issues

that will have to be addressed. In places, the haul roads in the current void are not wide enough for two 795Fs to pass one another, so it is quite likely that the trucks will have to be restricted to the higher levels of the push-backs and neighbouring extension. The new roads will need to be around 32 m wide in order to accommodate two of the new trucks, which means mine officials are also toying with the idea of purchasing an even bigger grader, such as the Cat 24M, possibly later this summer.

They are also considering a truck and trailer combination to speed up the transport of large machines around the mine. With a planned carrying capacity of 200 tonnes, this new machine should provide a solution to the difficult problem of transporting the large drilling rigs. Officials also plan to use this new vehicle to move broken-down trucks to the workshop by simply removing the trailer.

Finally, the mine superintendent secretly hopes to purchase the Komatsu PC8000 in the future. This enormous machine is Komatsu's largest excavator, weighing nearly 750 tonnes. 'A few years ago I had a go with one in the Canadian Oil Sands. It is a fantastic machine,' he says.

However they go about it, the underlying message is clear: the machines at Aitik are going to get even bigger as the mine grows into a world-class operation.

13.53. Truck tipping bodies in Europe do not get any larger than this one which belongs to a brand-new
Cat 795F AC truck – one of ten bought by the Swedish copper mine.

13.54. First revealed at the 2008 MINExpo in Las Vegas, New Boliden is one of the first customers for the Cat 795F AC.
Aitik is the only place in Europe that owns these huge haulers.

13.55. The Michelin tyres on the new electrically driven Cat truck are really massive and they have to be because when fully loaded, the new truck will carry well over 300 tonnes!

13.56. The 795F AC is made at Decatur in the US and it is a completely new truck for Cat, filling a gap between the 793 and 797. This machine is powered by a 16-cylinder Cat C175 engine that develops a mind-boggling 3,400 hp.

CHAPTER 14

Germany

Germany's role in the world of giant European earthmovers is highly varied and interesting. For starters, the country is the birthplace of the 750 tonne Komatsu PC8000, the huge components of which are made in the old Mannesmann Demag factory in Düsseldorf. Although eighty of these giant excavators have been produced, there are none working in Europe.

The super-sized 980 tonne Bucyrus RH400 – the world's largest and probably most expensive hydraulic excavator – is another massive machine which has its components made in Germany. The RH400 parts are made in a factory in Dortmund that was formerly owned by O&K and later by Terex Mining. Sadly, none of these phenomenal excavators work in Europe either; we have to make do with 'smaller' Bucyrus RH200s and Komatsu PC5500s.

These two factories are not the only earthmoving attractions in the country by a long shot. There are also a couple of other hugely popular destinations that regularly receive visitors from all over the world. The first destination, called Monster Park, lies in the centre of a triangle formed by the cities of Erfurt, Nürnberg and Würzburg, at Rattelsdorf in Bavaria.

The park is open Monday to Friday, March to October, and visitors to this extraordinary open-air theme park pay a small fee in order to tour the museum's fleet of 130 earthmovers from the past. This includes the chance to sit in the seat of a few rare O&K and Menck excavators, a mint-condition 60 tonne Poclain CK300 from 1981, Demag E30 from 1939, Clark-Michigan 475 wheeled loader and a Cat D9. The pride and joy of Monster Park is an 800 hp Demag H 185.

On weekends, Monster Park is transformed from a

14.1. Monster Park is the brainchild of construction contractor Gerhard Seibold. It is unique because it provides visitors with an opportunity to play with some seriously large earthmovers.

14.2. The Demag H 185 is the biggest excavator. Punters are allowed to sit in the cab and take photographs but they cannot play with it.

regular open-air museum to a test-driving site. Customers who pay a bit more are offered the mouth-watering chance to start the engines and move the controls of the active machine fleet, numbering around forty vehicles.

There is a similar initiative in England marked by the four Diggerland locations in Kent, Durham, Devon and Yorkshire. Also, there is the hint of another theme park being created in the Rhône Alps in France. For now, however, Monster Park is the only place in Europe where it is possible to pay to play with some seriously large earthmovers.

Access to the Monster Park machine fleet is split into three categories, which are organised by cost and size of machine. The first category offers an opportunity to work with mini excavators up to 6 tonnes and this is most popular with kids. The second class of machines are those with an operating weight of up to 30 tonnes and they are most popular overall. However, the most exciting and expensive experience is a spell on a monster truck, a monster wheeled loader or a monster excavator.

At one end of the sand pit, paying punters can start up and play with the bucket of a huge Komatsu WA800 wheeled loader, a rare 90 tonne capacity Faun rigid

dump truck and a couple of very old but very popular rope shovels. An 80 tonne H 65 front shovel excavator provides visitors with the thrill of a lifetime on the other side of the pit.

Spending just thirty minutes behind the controls of any one of these earthmoving icons costs the customers a hundred euros. Visitors keen to spend more time moving tonnes of earth have to be prepared to pay €320 for a two-hour stint. The most popular package for Monster Park's predominantly male visitors is a one-hour session costing €180.

The Monster Park site is not concerned with doing any real earthmoving work; instead, the experience is about digging up and burying holes in the 1.5 hectare play area. The opportunity to experience moving metal and lifting 15 tonnes of earth with the 3 m wide bucket of the H 65 is priceless to some customers.

There is no requirement to have a licence so anyone can join in the fun. The only stipulation is that the paying customers have to be over eighteen. Teenagers aged from fourteen to eighteen are not excluded but must be accompanied by an adult, while children, aged six and above, are restricted to the 6-tonners and must be supervised at all times.

14.3. The 80 tonne Demag H 65 is the largest excavator that visitors can use. The most popular session is a one-hour stint costing €180 in which people can start up the 500 hp engine and play with the front shovel bucket.

The Monster Park initiative is the brainchild of Gerhard Seibold, an enterprising businessman who heads up the Erdbau Seibold construction business. He conceived the unusual concept at the turn of the century towards the end of the post-reunification building boom. During this time, many earthmoving jobs were abandoned and, as a result, the machines stood idle. Gerhard soon realised that a considerable amount of people would be prepared to pay good money to sit on some of them. 'That's how the idea was born,' he says.

After buying a few machines, he allowed the public access to them on Saturdays – for a small fee of course. News quickly spread by word of mouth and within a few months he was regularly attracting fifty customers per day. In the end, exposure on national television was what really shot his business to fame.

Since then, Gerhard has never looked back. Many of his largest earthmovers are booked many weeks in advance, with punters keen for a spin on his private collection of metal monsters. Every Saturday, the site is crawling with machines chugging around the plot providing young and old alike with the time of their life. 'It is a boy's dream come true,' adds Gerhard.

His hobby has grown into a thriving money-making business, attracting as many as 10,000 visitors each year. There is also a more serious side to his business, however, because he also provides real training for people keen to obtain a licence to operate construction machinery for a living.

Skyscraper City

The second of Germany's major earthmoving attractions works further east in the huge lignite mines not far from the Dutch border. These are the largest mobile terrestrial machines on the planet and they are famous the world over. Motorists passing by on the autobahns in the Rhineland, west of Cologne, are often able to catch a glimpse of the tops of these towering mechanical monsters.

Although the site covers nearly 2,500 sq km and accounts for nearly three-quarters of the country's lignite or 'brown gold', many of the residents of Cologne and Düsseldorf are completely unaware that they are living next to the largest surface mines and machines in Europe. They are owned by RWE Power AG – the country's largest power supplier. Each year their unique fleet of bucket wheel excavators extract 100 million tonnes of Rhenish lignite from the region's three huge surface mines – Inden, Hambach and Garzweiler. Over ninety per cent of this lignite is used to generate electricity.

Of the three big sites, Garzweiler, which is located near the town of Grevenbroich just east of the A44, tends to attract the most attention. The lignite mine took its name from the 1,350 strong community unfortunate enough to be in the centre of the mining area.

Despite local resident and environmental protests, the site of the Garzweiler town was erased from the map between 1984 and 1989. The majority of its inhabitants were moved to a new location just north of the A46 called Neu-Garzweiler and all were well compensated. Incredibly, this was not the only village to be blitzed in this way. The settlements of Königshoven, Elfgen, Belmen, Stolzenberg and Priesterath have also been subjected to the same treatment.

This seems like a high price to pay, but alternative energy sources are thin on the ground in Germany. The country has very little gas or oil of its own and its nuclear power stations are due to be mothballed by 2025. Similarly, subsidies for hard coal mines are being wound down so many may close by 2018.

The plan is to use renewable forms of energy to generate thirty per cent of the country's electricity requirements by 2030. However, lignite and hard coal still play a crucial role generating one kilowatt hour of

14.4. The lignite mines are home to the largest earthmovers in the world – some stand over 100 m tall and weigh 13,500 tonnes.

electricity in four. The three German mines produce in excess of 10,000 megawatts.

Soft brown lignite is somewhere between peat and low-grade bitumen. The substance is to Germany what hard coal is to Britain and Spain, natural gas is to Holland, oil is to Saudi Arabia or the tar sands are to Canada. In other words, the substance is not only plentiful in supply, but relatively easy-pickings.

Lignite was discovered in the Rhineland in the 1700s and was used as a fuel for much of the following century. In 1910, the ability of the substance to generate electricity was realised and the area has since developed into the centre of a major coal-mining industry, with over three hundred years of estimated reserves.

The very first bucket wheel excavator arrived in the area in 1933 and by 1940 the twenty-three surface mines produced 60 million tonnes a year. In 1959, the former fifteen mining companies were amalgamated into Rheinbraun AG. In 1993, this company became RWE Power.

All three mines rely heavily on a fleet of bucket wheel excavators that are made by well-known names such as Orenstein & Koppel (O&K), MAN, Takraf and Krupp.

Even the 'baby' of the Garzweiler fleet – which was built in 1958 – weighs 3,500 tonnes, stands nearly 40 m tall, is over 125 m long and can dig 60,000 cu m a day. Then there are the four 8,000 tonne heavyweights, which are capable of shifting 110,000 cu m each during the same twenty-four hour period.

The fifth machine in the fleet of seven at Garzweiler is an even bigger machine known as 285. Ordered in 1970 and built by Krupp and O&K, this machine was delivered to the mine in 1976. The eighteen buckets fitted to the wheel of this 13,000 tonne giant are easily capable of teasing out 10,400 cu m an hour.

14.5. The massive bucket wheel excavator known as 288 weighs 13,500 tonnes and is 2 ½ football pitches long. It took four years to design and assemble this machine.

14.6. *The wheel on the front of 288 has a diameter of over 22 m and is nothing short of amazing. Every hour its eighteen buckets scoop up around 12,500 cu m of lignite or overburden.*

However, even the monstrous thirty-five-year-old 285 is dwarfed by a mega-structure known simply as 288. Towering nearly 100 m into the sky, 288 is a good 240 m walk from the tip of the bucket to the rear of the counterweight – roughly the length of 2 ½ football pitches.

The machine took four years to design and assemble and when it was ready for action in 1978, 288 superseded the crawler-transporter used to carry NASA's space shuttle in becoming the largest tracked vehicle in the world. The biggest man-made terrestrial mobile machine on the planet weighs an astonishing 13,500 tonnes and contains enough steel to build two Eiffel Towers. Connected to the front of the 70 m long girder boom is a massive rotating wheel that measures over 22 m across – equivalent to an eight-storey building.

Fitted with eighteen 6.5 cu m buckets, 288 can fill an average of forty-five buckets a minute. The massive machine operates by swinging gracefully from side to side across the face until a pass has been made, then the boom is lowered and the process repeated. Generally, six passes are required to scale the complete face.

Working relentlessly, not even stopping when the machine inches forward into the bench, the world's largest earthmover removes a mind-boggling 240,000 tonnes of overburden or lignite during three 8-hour shifts per day – enough to load a 30 km train or submerge a football stadium to a depth of 40 m.

There are no dump trucks at any of the three huge sites. Instead, 12,500 cu m of material an hour slides out of the buckets as they reach the highest point of the wheel's rotation. The contents fall onto a chute and then onto the boom conveyor. In the centre of the machine the material spills onto the conveyor of the telescopic connecting bridge. The length of this bridge can be altered by as much as 25 m to provide a bit of flexibility for the giant excavator to move forward.

At the end of the machine, a 1,800 tonne crawler-mounted discharge boom delivers the material into a hopper and the main pit conveyor, where it is whisked off at speeds approaching 30 km/hr to locally built power stations. Their lignite is in plentiful supply but the substance has a high moisture content – from forty to seventy per cent – which makes it uneconomical to move any great distance.

Even so, the three big mines contain over 240 km of conveyors. Garzweiler alone has 80 km and, of the 40 million tonnes of lignite mined, a large chunk speeds

14.7. The mega-structure is supported by twelve electrically powered steel tracks that measure 15 m long, over 3 m wide and 3 m high.

14.8. Bucket wheel excavators spew their contents onto conveyors.
Lignite is transported to the power stations at speeds approaching 30 km/hr.

14.9. Rhenish lignite has a high moisture content so it is uneconomic to transport it any great distance.
Lignite from Garzweiler is conveyed to Neurath power station in the background.

along the network of conveyor belts at the rate of 650 kg/sec to Frimmersdorf power station.

Garzweiler and Hambach are also linked by rail with the Niederaussem, Neurath, Frimmersdorf and Goldenberg power plants by 350 km of track. With thirty electric locomotives, seventeen diesel engines and 700 wagons, their combined efforts shift nearly 60 million tonnes of lignite each year.

Despite the immense size of 288, the machine only requires two operators – one to control the digging wheel from a cab close to the bucket wheel and a second to control discharge of lignite and overburden onto the main pit conveyor.

The bucket wheel boom forms part of the slewing upper carriage which is counterbalanced at the far end. Connecting the two is a network of wire ropes passed over the top of a pair of lofty steel lattice pylons. The act of lifting and lowering the bucket wheel boom is controlled by a hoist winch on the counterweight boom.

The operator at the digging end not only controls the boom's speed of swing, but also the depth of cut, wheel rotation speed and machine propulsion. While they are provided with a battery of CCTV images and graphical information on system performance, the operators can also rely on radio as a way to stay in touch with the wheeled dozer or loader operator far below.

The two operators are aided by one maintenance mechanic as well as a ground man. A crucial part of this person's job, as well as staying in close contact with the two operators at all times, is to ensure that there are no obstructions when the machine moves and that it does not run over its electric supply cable.

The importance of this task is best highlighted by an incident that happened a few years ago when one of the bucket wheel excavators scooped up a Cat D8 dozer – the pictures of which quickly circulated the internet. The excavator operator did not even realise what had happened until the dozer was at the top of the wheel. Subsequently, a huge crane was needed to remove it.

Bucket wheel excavators can only be used in rare and very special geological conditions and will not work where rocks or boulders are present. They can easily

14.11. The second operator sits in an elevated position in the mid-section. He controls the discharge boom, hopper movements and ensures a smooth supply of material to the main pit conveyors.

14.12. The boom is suspended by a series of hoist ropes that pass over the top of a pair of steel lattice pylons to the counterweight. It can tackle a face height of 51 m when extended to full height.

14.13. The geology in the area is perfect for the spinning wheel because both the lignite and the overburden are soft. Each year, the massive machines remove 100 million tonnes of lignite and between 500 million and 600 million tonnes of overburden.

bring the world's largest earthmover to a standstill. The geology has been kind in this particular area of Germany where the lignite – as well as the clay, sand and gravel that cover it – is soft. These conditions are perfect for the Garzweiler bucket wheel excavator extraction fleet but not suitable for deep mining activities, such as those in the Ruhr coal-mining area.

Unlike rope shovels, draglines or excavators, bucket wheel excavators do not need to stop between cycles. Instead, they offer a continuous cycle, shifting colossal volumes, which is arguably the most efficient way to dig and the least punishing in terms of stress loads.

There are no jerky movements as the electrically powered boom glides gracefully from side to side. The phenomenal digger hides thirty-two electric motors and over 160 km of wires and cables – all designed and installed by Siemens.

Travelling at a maximum speed of just 600 m/hr, each of the twelve massive steel tracks is over 15 m long, nearly 3 m wide and 3 m high. Covering a total width of 46 m, 288 puts down an enormous 660 sq m footprint, helping to reduce the ground pressure to just over 1 kg per sq cm.

14.14. With the exception of routine maintenance, there is no rest for the giant machines. Working round-the-clock, four bucket wheel excavators are able to scoop up 110,000 cu m every twenty-four hours.

14.15. Snow turns the lignite mines into winter wonderlands but otherwise the giant machines are unaffected.

The company hopes to keep the digging wheel of their monster miner spinning for at least another sixty years – not unreasonable considering that it would cost a mind-boggling €100 million to replace it.

The largest earthmover in the world was originally built to remove overburden at Hambach. When that job that was completed in February 2001, the massive machine was moved 30 km to Garzweiler.

During its epic journey from one mine to the other, it kept a team of seventy busy for three weeks. In that time, it crossed several roads (including the A61), the Erft river and a rail line. It even required the removal of several power lines. The biggest challenge was the river because steel pipes had to be covered by rocks and gravel to create a smooth surface to travel across.

14.16. The largest mobile terrestrial machine created quite a stir when it was moved from Hambach to Garzweiler in 2001. It took three weeks to complete the epic 30 km journey.

14.17. A decade ago, visitors came from all over Germany to see 288 trundle slowly to its new home.

14.18. Over the years, many of the large bucket wheel excavators have been moved.
Here, one of the mobile monsters crosses the Autobahn 61.

14.19. *Many men and machines are required to prepare a level surface for bucket wheel excavators to travel on.*

The move cost an estimated €7 million, but this was still far cheaper than dismantling the massive machine and transporting it piece by piece. Since then, this method of transportation has been adopted by several other mines.

Garzweiler II

When the lignite outputs at Garzweiler were exhausted, some of the mega machines continued eating their way west into a new home. The new space, called Garzweiler II, covers an area of 48 sq km. It is estimated to hide 1.3 billion tonnes of lignite – forty per cent of the known reserves in the Rhineland region – which is probably enough to last until 2045.

In the 1995 elections of the state of North-Rhine Westphalia, the new operation at Garzweiler II was high on the agenda. The environmental discussion as to the merits of mining for lignite continues even today, but fifteen years ago there were many other issues for the politicians and activists to contend with. The A44, a road directly in the path of the bucket wheel excavators, was just one such issue. The village of Otzenrath and many others after it also presented the massive machines with more obstacles.

The Green Party won the confidence and support of many voters with a pledge to stop the advancement of the Garzweiler II site. However, the planned mining extension went ahead anyway, a decision which not only secured some 12,000 local jobs but another 40,000 in ancillary industries. With approval granted, the A44 between Düsseldorf and Aachen was permanently closed in October 2005 and traffic was re-routed onto the parallel A61, which was widened in advance from two to three lanes.

Then a section of the closed A44 was demolished to allow the big machines to continue mining into Garzweiler II. The A61 will face the same fate in 2017, when traffic will be diverted onto a newly built A44. By 2035, the plan will eventually come full-circle when the A44 will be once again scrapped and traffic re-diverted onto a re-built A61.

In 2006, work started on two new state-of-the-art power stations, called Neurath and Niederaussem, which were to become the largest of their type in the world. Located north-east of Frimmersdorf, these power stations cost an estimated €2.4 billion. They were exceptional because both use pre-drying techniques to reduce the water content and boost efficiency to around forty-five per cent – a ten per cent improvement on the older power stations that is actually not far behind the efficiency of hard coal. In addition, to produce the same amount of electricity, each one of these state-of-the-art facilities also reduces CO_2 output by between twenty-five and thirty per cent.

As can be seen, it is not just the machines that occasionally move; sometimes complete villages need to be relocated in order to continue excavating lignite. It could be argued that RWE Power is not only an energy supplier but is also in the relocation business. Since the end of World War II, it is estimated that lignite mining activities have not only re-routed roads but have also destroyed close to fifty villages, causing the forced re-housing of 35,000 people.

14.20. Seven of the bucket wheel excavators owned by RWE Power are currently digging their way into Garzweiler II. By the time it is finished (around 2045), it will have yielded 1.3 billion tonnes of lignite.

14.21. Lignite mining activities have led to the destruction of an estimated fifty villages and the forced re-housing of 35,000 people since the 1950s. Many more are estimated to follow before lignite reserves are exhausted.

In the path of their large machines in Garzweiler II, the village of Otzenrath has already been erased from the map and its residents resettled at New Otzenrath, north of the A46 near Jüchen-Hochneukirch. The relocation of the towns of Holz and Spenrach is also nearing completion. Pesch and Borschemich will be next. When the time comes for the giant fleet to march across the A61 in 2017, six other villages – Lützerath, Immerath, Kuckum, Keyenberg, Oberwestrich and Unterwestrich – are ultimately doomed to extinction.

Residents in Berverath and Holzweiler will eventually need to vacate their homes, but they still have some time to prepare because the lignite under their homes is not needed until 2030. By the time that extraction winds down, the moving Garzweiler II site will have cleared a dozen villages and caused the forced resettlement of over 7,600 people.

Heaps of Water

The lignite at Garzweiler II is contained in three major bands, the first layer of which ranges in depth from 2 to 20 m and is covered by a thin band of topsoil and 40 m of sand. The second layer measures between 20 and 30 m and can be found buried beneath 40 m of overburden. The third band is approximately 20 million years old. The depth of this ancient band of lignite is 30 m and it ends at a depth of 210 m.

Before crews and machines can even consider working at a depth near 210 m, ground water needs to be removed. An extensive network of pumps and pipes extract a staggering 600 million cu m of water every year from the deep seams at Garzweiler II.

Some of this water is used as drinking water, while large volumes are also required to cool power stations.

14.22. The electric spreaders form an important part of the mining operation. The largest of six at Garzweiler weighs nearly 4,500 tonnes and is quite capable of spreading 240,000 tonnes during a single 24-hour day.

14.23. The Caterpillar dozers have the never-ending task of levelling the spoils from the spreaders.

Much of the remainder is stored and will be pumped back into the land when mining has been completed and the land is restored to farmland.

Wandering slowly across the German countryside eating up the lignite is just one part of the operation because the void left when the bucket wheel excavators have done their work needs backfilling by smaller machines known as spreaders.

All overburden extracted at Garzweiler II will be used to backfill Garzweiler I. Before the overburden is laid, it is treated with limestone to prevent acidification and then covered with a layer of topsoil to bring the land back to as natural a state as possible. Within seven years the site should be fit for farming and forestry activities, but it will take another thirteen years before the land is stable enough to support buildings.

14.24. When the machines have finished slowly eating their way across the German countryside the area is restored. Eventually it will be used for farming or forestry.

14.25. Paffendorf castle at Bergheim is a great place to learn more about lignite mining. It houses a permanent exhibition, some interesting archaeological finds and a couple of buckets from one of the giant machines.

Visitors' Paradise

The 400-year-old Paffendorf castle at Bergheim is a great place to begin learning more about this massive industry. The castle is open during weekends and public holidays and it houses a permanent exhibition – some of it in English – on the origin and exploitation of Rhenish lignite, including archaeological finds and even a couple of the buckets from one of the giant bucket wheel excavators that works at the mine.

There is also an exhibition in the Coal Innovation Centre at the Niederaussem power station as well as an information centre at the Weisweiler power station. Both are open on Sundays and can be visited free of charge.

For those keen to see the world's largest earthmovers in the flesh, the opportunities do not end with museums and information centres. Visitors and enthusiasts are invited to attend one of RWE Power's four hugely-popular – not to mention free – open-day Sundays. These events are generally held in May, June, August and September; they are organised on a first-come, first-serve basis with each one attracting nearly 2,000 people.

Buses running from a parking area near the tennis hall at Bedburg-Kaster provide easy transport to RWE

Power's sites. The queues can be long so it pays to arrive early. The first bus leaves at 10am and the last leaves at 4pm. The open-day tours last about one hour and one of the highlights is a close encounter with the largest earthmovers on the planet – albeit from within the confines of the bus.

Alternatively, the company also offers guided technical tours for groups with their own bus; these last 2 ½ hours. Every year, this popular option allows 40,000 visitors to get up close to these spectacular beasts.

Together with the 8,000 people who join the four annual Sunday open days and the 50,000 who visit the power stations, the mining operation in this region of Germany welcomes around 100,000 visitors each year – sometimes from as far away as Taiwan, America and Australia.

The tour of what is sometimes referred to as Europe's Grand Canyon, leaves many people with mixed emotions. On the one hand, people are confronted with man's changing of the landscape and with the destruction of communities once home to generations of people. On the other hand, the grand scale of the operation, the size of the equipment and the deep gorges they leave in their wake creates a sense of awe.

The fact that RWE Power allows visitors inside their mines and power stations in the first place is part of a policy of transparency. Most of the world's large surface mines are not keen on visitors because they are afraid of the adverse publicity they might attract. The German energy giant is unique in that it makes a point of stressing that it has nothing to hide.

In the spirit of transparency, RWE Power has even created a map known as the energy trail, which is available at Paffendorf Castle and can be downloaded from the Internet. The route provides clearly marked places of interest for cyclists or motorists eager to learn more about lignite and how it is extracted and used for electricity. It also describes how the land is restored afterwards. They have created viewing areas that provide a stunning peek into the bowels of the big mines. It is important to remember, however, that the shape of the mine changes so quickly that it really pays to check which routes are still open before attempting to follow them.

At Inden, visitors can take a first-hand look at results of the reclamation work and follow current mining activities from one of three viewing points. Not far away at Hambach, visitors can see a number of local

14.26. RWE Power encourages visitors by organising guided tours through the three mines. They have also created a number of accessible viewing points like this one at Hambach.

landmarks formed by the spoils of earlier mining activities. These include the Sophienhöhe, a 290 m high heap of overburden that dates back to when the mining activities started in 1978.

East of Jüllich and north of Hambach, the Sophienhöhe lies in the middle of what used to be the site of the 5,500 hectare Hambacher forest. Today, this area is something of a recreation ground, providing 70 km of trails for

14.27. This site is Inden, one of three big surface mines in the Rhenish lignite field belonging to RWE Power.

walking, mountain biking and horse riding, not to mention some spectacular views of the big machines working not far away.

Local travel agencies also cash in on the popularity of the area's lignite mining. The Rhine-Erft Tourism Association offers the exciting prospect of a three-hour coach tour of castles and bucket wheel excavators. The tour includes a visit to the 'lignite experience' at Paffendorf Castle and a stop at the viewing point at Hambach after lunch, followed by a tour of some of the villages destined for destruction and some of the newly built communities.

With all the activities and avenues open to the earthmoving enthusiast, it might be difficult to accept that eventually the lignite mining in Germany will have to come to an end. Fortunately, there is still plenty of time left to visit the largest earthmovers in world. Although some of the machines are now over fifty years old, they should survive until the Garzweiler activities finish, which will be around 2045. Once the restoration process is complete, part of the sprawling site will be transformed into a 20 hectare lake and the massive machines will have to be relocated.

The burning question is how long will the world's largest earthmovers really be able to continue their seemingly unstoppable and relentless quest to mop up the rest of the lignite lying below the surface? With renewable energy sources providing a more favourable solution to uprooting roads and villages, the digging wheels of these feats of human engineering may ultimately grind to a standstill.

At present, Germany is in the midst of plans to increase the share of renewable energy from a current twenty to thirty per cent by 2020, but the country is still very much reliant on conventional sources to provide seventy per cent of its energy requirement. For now, at least, Germany needs its lignite and the monster machines used to extract it.

Made in Germany

Some of the world's largest mining shovels are made at the Komatsu Mining facility in Düsseldorf, not far away from the massive lignite mines of the Rhineland.

The current PC Super Shovel series can trace their roots to a factory located near Duisburg where Carlshütte built the first electrically powered rope

14.28. The B 504 weighs just over 12 tonnes and was the first diesel-powered all-hydraulic excavator in the world. Made in 1954, this fine example of these early machines stands just inside the main entrance of the factory owned by Komatsu Mining Germany.

shovel. The factory was acquired by Demag (Deutsche Maschinenfabrik AG) in 1937 and immediately construction began on a dedicated excavator plant. In 1949, it produced the first B 300 and B 400 rope shovels and was known as the Benrather Baggerfabrik.

In 1954 the factory built the first B 504 which weighed just over 12 tonnes. It was the world's first diesel-powered completely hydraulic excavator with a 42 hp air-cooler Deutz engine. A fine example of one of these early machines stands just inside the main entrance to the present-day factory.

Eighteen years after the introduction of the B 504, the same factory produced the first large diesel-powered hydraulic mining shovel in the world. It was released in 1972 and called the H 101; the 100 tonne machine supported a 5 cu m bucket.

In 1980, the business side of things changed when Demag was taken over by the Mannesmann Group; the name was changed to Mannesmann Demag. During the decade that followed, the excavators continued to evolve, not only in sophistication, but also in capacity. Costs per tonne were becoming increasingly important and bigger buckets helped to improve efficiency.

In 1996, the Japanese company Komatsu became involved when it bought fifty per cent of Mannesmann Demag. Three years later Komatsu completed the process when it bought the remaining shares and changed the name of the business to Komatsu Mining Germany. For the next few years the new owners took a long, hard look at the four mining excavators it produced and devised a plan to position them in the market. The aim was to make Komatsu the Mercedes of the shovels.

PC3000

The origins of the popular present-day 260 tonne mining shovel go back as far as 1978 with the launch of the 12 cu m Demag H 185. From 1990 to 1991, the H 185 evolved into the H 185S, which was a 14 cu m machine particularly popular in the UK.

In 1998, Komatsu owned half of the company and the takeover prompted a change in the finer details of the old H 185S. Its successor was the 14 cu m H 255S. This model was eventually converted to the 14 cu m PC3000-1, then, in 2002, it was tweaked and rebranded as the 15 cu m PC3000. Whatever the name, this model is another extremely popular earthmoving machine in the UK.

Of the 178 PC3000 excavators produced in Germany so far, a couple of dozen can be found in Britain. The German factory is able to eliminate a bottleneck in production by utilising the Kanazwa plant in Japan as an additional manufacturing facility. Seventy machines have been made in Japan to date and they have been delivered all around the globe.

PC4000

The PC4000 originates from the single-engine H 241 which was produced in 1983. The H 241 was an even bigger beast with an operating weight nearing the 300 tonne mark and bucket capacities up to 14 cu m. In 1986, the H 285S replaced the H 241 as a 310 tonne giant carrying a 19 cu m bucket.

In 2001, the 380 tonne PC4000 was unveiled at the Bauma show in Munich. Replacing the H 285S, the newcomer weighed over 390 tonnes and carried a 22 cu m bucket. Eighty-two of these excavators have been produced in the last ten years. In fact, the two that were the very first pair in Europe – PC4000-6 electric front shovel versions – are now working at a copper mine in Serbia.

PC5500

In 1995, the German factory created the H 455S – a 500 tonne excavator with a 25 cu m bucket. The evolution of the PC5500 is marked by the growth in capacity. The PC5500 was originally equipped with a 26 cu m bucket. In 2006, it rose again to 28 cu m, supported by the 530 tonne PC5500-6. Finally, in 2008, the Stage IIIa version of the excavator was debuted at 29 cu m at the Las Vegas MINExpo.

PC8000

The PC8000 is not only the biggest of the PC range of Komatsu excavators it is also a hugely successful model in its own right. The formidable machine we know today actually started out in 1986 as a Mannesmann Demag H 485 – an excavator that was coupled to a massive 23 cu m bucket. This machine broke the record for the largest operating weight – over 500 tonnes – and became the world's largest hydraulic excavator. It was offered with a single Siemens electric motor capable of producing 1,650 kilowatts (2,212 hp), or a single 16-cylinder MTU diesel engine capable of producing 2,105 hp (1,570 kilowatts).

The very first H 485 in the world was delivered to a Scottish company called Coal Contractors. The company went for the diesel engine option. Eventually, the excavator was shipped to Canada and it is assumed that it has been scrapped or broken up for spare parts.

Even though these excavators are now getting old, a few of these impressive diggers still provide their owners with reliable service after tens of thousands of operating hours. In America, an excavator with the serial number 12024 had an unusual experience. After 90,000 hours, it had its two diesel engines replaced by electric-drive to give it a new lease of life. The owner is confident that the old PC8000 will work for another 60,000 hours!

In 1989, one of the first commercial versions of this machine was commissioned in Australia at Similco Mine's Copper Mountain operation. This one was nicknamed 'Doris'. Although it was delivered with a single MTU engine, this was later swapped for two Cummins power packs pumping out over 3,000 hp. It wielded a 26 cu m bucket.

From 1989 to 1991 Mannesmann Demag added an electric version of the H 485. One of the first machines to excavate in hard rock worked at theSwedish Aitik copper mine. This machine was specially adapted for working north of the Artic Circle and it was so successful that it was joined a couple of years later by a second. Sadly, the last of the pair was scrapped three years ago.

In 1991–2, the H 485 developed into the H 485S, which was capable of carrying a 33 to 36 cu m bucket. Two years later, the 35 cu m H 485SP followed; this machine had a service weight of 585 tonnes. This excavator paved the way for the 36 cu m H 655S during the period 1997 to 1999. The H 655S is a 685 tonne machine that is so special that it is only found in the Canadian Oil Sands.

The launch of this extreme excavator saw the single Siemens electric motor ousted for two ABB ones, which combine to generate 2,782 kilowatts (3,620 hp). The advantage of two power trains is that they allow for more flexibility and provide an affordable way to increase the power. The new engines also brought additional safety allowing the shovel to be moved if one power train suffered a breakdown.

At the same time the Cummins diesel engines also disappeared in favour of Cat as the new power source for the PC8000. Cat engines could provide 3,730 hp (2,782 kilowatts) of power to the massive machine. In

2004, while the engine options remained unchanged, the bucket capacity rose again; the 700 tonne machine, known as the PC8000-1 came equipped with a 38 cu m bucket.

Cold Climates

The next milestone in the development of this industry-leading machine came in 2006. The company was commissioned to produce a unit fit to work in the harsh Siberian climate where winter temperatures regularly plummet to minus fifty degrees Celsius.

The standard PC8000 machine can operate in temperatures as low as minus twenty-five degrees Celsius. The excavator was fitted with pre-heating systems for the engine, cab and hydraulic system, while special steel plates and different seals were also fitted to help the excavator cope with digging in hard rock permafrost. Large mining excavators do not just work in the cold but also at high altitudes so Komatsu made a few adjustments to a couple of PC8000s to allow them to work at an altitude of 5,200 m.

In 2007, the service weight received another boost to 737 tonnes and the bucket capacity rose to 42 cu m to allow it to fill the newly launched 290 tonne capacity Komatsu 930E dump truck in just four passes.

After the boost in weight and bucket size, Komatsu found that the PC8000 needed more power. The output was increased to 2,900 kilowatts (3,888 hp) on the electric motors, while the Cat engines were replaced by Komatsu engines capable of kicking out 4,023 hp (3,000 kilowatts). The result provides a phenomenal digging force of around 236 metric tonnes, which is roughly equivalent to a force of 40 tonnes per tooth as the bucket is forced into the face. In other words the force of the PC8000 is equivalent to placing the force of a fully loaded truck onto the surface of the bucket teeth tips instead of across the twelve wheels!

All the super-sized components of all four PC Super Shovels, including the PC8000, are made and then assembled in Düsseldorf. Understandably, it is difficult to appreciate the size of these machines, especially the giant PC8000. Each track of Komatsu's top-end excavator is over 10 m in length and the machine stands almost 10 m tall. Then there is the engine housing, which covers a staggering 90 sq m – roughly the size of a small flat! The PC8000 is available as a front shovel or backhoe. On a good dig the 42 cu m bucket is easily capable of shifting 7,000 tonnes an hour.

14.29. Inside the factory, the serious business of joining the two halves of the huge 42 cu m bucket is underway.

14.30. Many of the components of the PC8000 are so large they need to go out on their own truck.

14.31. Komatsu's PC8000 weighs 750 tonnes. It is a successful machine, over eighty of which have been made so far.

All machines built at the German Komatsu factory are so large that they are not married to their undercarriage and digging equipment until they arrive at their destination. Even the largest bucket is too big to transport as a single lump so instead the two halves are joined together at the end of the journey. They are, however, thoroughly tested as a complete unit during the final assembly process at the factory.

It is sometimes difficult to comprehend that while this massive miner is just too large for Europe some of the world's largest mines view the PC8000 as an auxiliary machine, playing second fiddle to much larger electric rope shovels!

Although it is unlikely that a PC8000 will ever work in Europe nothing is impossible. The best place to see them is in South America, which contains over thirty machines – currently the highest population of the excavator anywhere in the world.

The Present

When the older Mannesmann Demag and more recent Komatsu mining shovels are added together, the German factory reckons that it has shipped out over 700 active mining shovels since the 1980s, with two-thirds

of them still working today. The largest population – over 150 – work in Latin America.

It is not just their mining excavator family that has changed over the years. So too have the manufacturing and assembly facilities at Düsseldorf. Recent investments mean that the company can make fifty mining shovels a year. Komatsu can also rely on the plant in Japan to assist with the production of PC3000s and PC4000s.

Standards are exceptionally high. For example, in Germany, welders are officially required to pass a test every two years. Komatsu goes one step further by requiring that every one of their welders takes a test every year. There is good reason for this attention to detail; large mining shovels contain several miles of welding, most of which is done by hand. There is no room for error because all welds are ultra-sonic tested. Each month the company allows only 9 m of welding to be re-done. In what is internally know as the Komatsu welders' Olympic Games, every welder has chance to take part in a global test of skill; it may come as no surprise that Komatsu Mining Germany welders are continuously on the podium.

Other strict quality control measures apply to the paint used to finish their mining shovels. Not only must it be able to withstand the world's hottest and coldest temperatures, but also rapid temperature fluctuations. For this reason, they subject it to quite severe tests where they cool it down to minus fifty degrees Celsius for an hour, then raise the temperature quickly to plus fifty degrees Celsius for another hour. This procedure is repeated three times and the paint must not crack.

There is little that can destroy a PC8000. The greatest risk is a fire in the engine department and Komatsu has developed an agreeable solution to this problem as well. To reduce the possibility of fire, the engines and hydraulic pumps are separated by a protection wall. Hot surfaces are covered, while holes in the massive counterweight allow water to be forced into its belly in the unlikely event of a fire.

World's Largest Excavator

Another bunch of the world's heaviest miners are made not far away in a factory in Dortmund. The factory was founded by the famous duo Benno Orenstein and Arthur Koppel, better known as O&K. Initially, O&K specialised in railway machinery, but when the premises were practically blitzed during World War II, the site had to be rebuilt. The very first O&K excavator – the RH5 – was designed and built in the early 1960s.

The RH5 had a service weight of 18 tonnes, was powered by a 60 hp engine and its digging end supported a bucket with a mere 0.5 cu m capacity. Today one of these iconic excavators is parked inside the main entrance of the Dortmund factory.

The O&K brand rapidly developed into one of the biggest names in the business, making tens of thousands of excavators, including hugely popular models like the RH40, RH120 and RH200.

Never in their wildest dreams could the famous duo have predicted that their business would ultimately go on to supply a hundred mining shovels a year. The active machine park – starting with the RH40 upwards – is estimated to be in excess of 1,500 excavators worldwide.

In 1998, the company was acquired by the Terex Corporation and the name was promptly changed to O&K Mining and later to Terex O&K. This takeover also saw the change from the traditional red O&K livery to the new white with red trim Terex livery.

In 2009, Terex sold the mining division of Terex O&K to Bucyrus – an operation that could be short-lived. In late 2010, Caterpillar lodged a multi-billion dollar bid to buy

14.32. This O&K RH5 weighs just 18 tonnes and is parked inside the main entrance of the Dortmund factory currently owned by Bucyrus.

the Bucyrus company. Poised to take control, it remains to be seen what plans Cat may have in store but, for now, at least Bucyrus continues to offer former O&K machines, such as the 105 tonne RH40.

Other large excavators built in Dortmund include the 170 tonne RH90, the 285 tonne RH120, the 397 tonne RH170 and the 530 tonne RH200. At the 2004 Bauma show, Terex added a new model to the range – the 568 tonne RH340. Even today, this is the only excavator that has a standard bucket – at 34 cu m – that matches its model number.

RH400

All these machines are completely eclipsed by the gigantic 980 tonne RH400, which is the largest hydraulic excavator in the world. Unveiled in prototype form in 1997 and officially launched at the 1999 Bauma show, the RH400 was developed by Terex Mining.

In the mid-1990s, the biggest hydraulic excavator carried a 35 cu m bucket but potential customers wanted an even bigger bucket to provide a four-pass match to new trucks that had a higher capacity of 320 tonnes. In 1997, a year and a half after starting the project, the massive excavator was ready to start work in the oil sands at Fort McMurray in northern Alberta.

Power to the industry-leading excavator comes from a pair of Cummins K 2000 E diesel engines – the largest available from Cummins at the time. The engine was capable of delivering a total of 3,350 hp and there was good reason for all the power because up front the excavator carried a 42.5 cu m bucket. Production tests confirmed outputs of 9,000 tonnes an hour which marked a record for hydraulic shovels. Six months later, the first RH200 at Fort McMurray was joined by a second machine that was fitted with two even larger 60 litre Cummins QSK 60 engines.

14.33. This picture really emphasises the huge size of the components of the massive Bucyrus RH excavators.

14.34. What a thrill it must have been for these German school kids to clamber inside the mighty bucket of the largest hydraulic excavator in the world before it was pulled apart for the long journey to Canada.

Nearly two years of field experience and detailed monitoring of the first two prototypes resulted in a number of changes to the massive excavators. Commissioned in 2000, the next two RH400s carried a longer boom and stick which gave them more reach but required the lengthening of the undercarriage. Both needed a heavier counterweight and on top of that they were given a bit more power in the form of two Cat engines pumping out 4,400 hp. At this point, the RH400 had an operating weight approaching 1,000 tonnes and was easily capable of lifting 85 to 100 tonnes at a time.

The four diesel-powered excavators were followed by an electrically powered fifth RH400 in 2000 – the first with the Terex livery. The fifth excavator was bought by a US coal mine and fitted with two electric motors from the RH200. It was followed a year later by a second, also in the red-and-white trim livery, which went to the Canadian oil sands.

14.35. The RH400 is also available as an electrically powered excavator. Made in 2000, the first one was fitted with two electric motors from the RH200.

Over the course of the years, more diesel-powered versions have followed to customers in South Africa, Australia and both North and South America. The RH400 is available with two engine options – Cat or Cummins – both of which generate a maximum output of 4,500 hp. They are both fed from a 16,000 litre fuel tank.

Although it is highly unlikely that one of these giant earthmovers will ever work in Europe, we can take comfort in the knowledge that at least the largest hydraulic excavator is European-made.

14.36. Pictured here loading Caterpillar's largest dump truck, the RH400 is a colossal excavator. Tipping the scales at 980 tonnes, its bucket can lift anything from 85 to 100 tonnes at a time.

14.37. Bucyrus ownership has not only brought a change of livery, but also a change in the model numbering with the former space between the RH and 400 now closed.

The Dutch ENCI quarry recently took delivery of a new Cat 988H. Their older 988F has been retired.

I always knew a couple of things were likely to change some time between completing the final chapter and publication of this book – hence this postscript.

In France, Rio Tinto Minerals has just sold the Luzenac talc mine to Imerys. Let's hope the new owner will allow tourists to continue visiting the stunning mountain-top mine in the Pyrenees.

In the Netherlands, staff at the ENCI limestone quarry near Maastricht recently welcomed two new arrivals. The first, a new Cat 988H wheeled loader, was bought to replace an exhausted Cat 988F, while an equally new Cat 775F dump truck will soon be ready to take the place of one of the quarry's older haulers.

Their only remaining Liebherr R 994 excavator remains a cause for concern and will certainly not survive until quarrying activities end in 2018. However, ENCI are doing everything possible to keep it going for as long as possible. It will soon get a new set of tracks gleaned from an old excavator at a limestone quarry just across the border in Belgium.

The most serious issue for this old Liebherr is that it has a crack in the slew ring casing that will eventually end its working life. Employees at the quarry believe

that it will survive another 4,000 hours – roughly two more years – at best.

So the search continues for a successor – new or used. There have even been hints that the quarry might hire a contractor to tackle the softer limestone at the top of the quarry.

Grensmaas Revisited

At the Grensmaas project, to the north of Maastricht, Dutch contractor Janssen has swapped the long-reach equipment on a Cat 385C for a short arm. This new arrangement has been coupled with an 8 cu m bucket and is capable of loading as much as 1,600 to 1,700 tonnes an hour.

Elsewhere, two of the Cat 385s that were on gravel-extraction duty were recently replaced by a pair of brand-new 385Cs. One machine came from Van Oord and the other from Van den Biggelaar; both are well-known Dutch contractors. The excavators are supported by specially designed 5.5 metre square undercarriages. They are heavily beefed up to carry larger hydraulic cylinders with a 22 m boom and stick coupled to a 5.5 cu m bucket.

One of two practically identical Cat 385Cs at the Grensmaas project in the Netherlands.

There are thought to be just half a dozen of these specials worldwide. The pair at Grensmaas were specifically modified for the gravel job. Each one excavates around 11 tonnes a time – as much as 1,000 tonnes an hour.

Maasvlakte 2

Since writing the Dutch chapter I have been fortunate to spend more time at the Maasvlakte 2 project in the Port of Rotterdam. The massive land expansion project continues to enjoy unprecedented attention; this year alone, an estimated 115,000 visitors are forecast to tour the facility, making it easily the most visited earthmoving project in Europe. To top it all, the FutureLand information centre has recently recorded its 200,000th visitor.

The one-hour journey across the new land takes visitors right past the first of the new quay walls, which are now in an advanced stage of construction. While the tour stays well clear of the Blockbuster building the new hard sea defence, there is plenty of information about the project available in FutureLand.

Even though visitors have to keep a distance from the phenomenal machine, the Blockbuster continues to enjoy most of the attention on site. Most impressive is its ability to place massive 40- to 45-tonne concrete blocks with stunning accuracy. The Blockbuster is fed with blocks from the original hard sea defence, and – at the time of writing – the colossal excavator that provides the blocks for the Blockbuster was busy extracting the final blocks from the waters right in front of the information centre.

This monstrous machine is a Liebherr P 995 fitted to a pontoon called Wodan. The digging arm fitted to the P 995 carries a specially made grab to pluck blocks from the murky waters. The only land-based version of this excavator – the R 995 – is featured in the Spanish chapter. The P (pontoon) version carries the top half of the same excavator.

Not far away, an identical Liebherr P 995, called the Nordic Giant, is on dredging duty. This phenomenal underwater digger is captured on film, along with many others, in the DVD *Massive Backhoe Dredgers* – one of the 'Massive Machines' DVDs produced by Old Pond.

The top half of a massive Liebherr P 995 excavator digs up concrete blocks at the Maasvlakte 2 project in the Port of Rotterdam.

Japanese Miners

The Dutch may be masters of winning land from the sea, but the main reason for this Postscript concerns the recent developments that have occurred in South Wales where quarries and mines rely heavily on Japanese-made mining excavators from Hitachi and Komatsu.

Little could I have realised that Japan would have to feature so strongly due to the devastating effects of the tsunami that hit the country's north-eastern coast and the earthquakes that followed. Of course, any immediate thoughts were with the tens of thousands of people robbed of their homes, possessions and loved ones.

The natural disaster had horrible consequences for

Nant Helen is one of three Celtic Energy surface coal mines in South Wales.

Japan's nuclear power sector and the full effects of the radiation that continues to leak from the Fukushima nuclear power station could still take many years to surface.

The problems with the reactors at Fukushima have added another dent to the worldwide image of the nuclear power industry because once again this latest disaster raises important human safety issues. Coincidentally, the disaster in Japan occurred only a few weeks before the 25th anniversary of the Chernobyl explosion, which happened on 26 April 1986.

The Chernobyl disaster clearly indicates that nuclear problems will not disappear overnight. Even twenty-five years later there is still a 30-km exclusion zone around the complex and perhaps more sinister still is the concrete sarcophagus encasing the tonnes of radioactive material which is at the end of its life. The concrete is now showing signs of cracking and the Ukrainian government has been forced to seek help from the EU, over €700 million to be specific, to build a new casing to slide over the top of the old one.

Unfortunately, it is Japan's turn to deal with the effects of the tsunami and the earthquakes that happened in March, and they are not alone. Although this latest incident occurred on the other side of the world, the repercussions are already being felt in Europe.

In light of what has happened in Japan, rather than invest an estimated €5 to 6 billion to renew each of its five nuclear power stations, Switzerland plans to spend the money on safer and more renewable forms of wind and water-generated power.

However, the biggest blow to the nuclear sector comes from Germany, which has announced plans to shut all but one of its seventeen power stations by 2020. Forced by public opinion to make a U-turn in energy policy, Germany's new plan is to pump billions of euros into wind and solar power. They will need plenty of both because nuclear power currently accounts for nearly a quarter of the country's electricity requirements.

This prompts the question: will Germany be able to fill the gap produced by abandoning nuclear power with

After nearly 30,000 hours of operation, Nant Helen's PC3000-1 is up for sale. The excavator still has a good residual value.

wind and solar energy alone? It is highly unlikely given the timescale.

Either way, the government's decision will be welcomed by Germany's lignite mines because they will probably be needed for longer than was originally planned. If other European countries decide to follow in the footsteps of Switzerland and Germany, then this crisis of faith in nuclear power could very well provide European coal mines (and their big machines) with an unexpected lifeline.

While many of the large hydraulic excavators featured in this book were made in Europe by Komatsu, Liebherr and Bucyrus, Komatsu also manufactures quite a few mining machines in Japan, as does Hitachi. Some of the factories of both these popular brands were disrupted by the tsunami and the earthquakes but work to restore buildings and production facilities started immediately and in a matter of weeks some factories were back to manufacturing at full strength. Unfortunately, others still need a bit more time to resume normal production and clear the backlog of orders.

Closer to home, the Welsh chapter of this book features a pair of Japanese-made 200 tonne Hitachi EX1900

excavators belonging to Celtic Energy. When new, the first EX1900 in Britain – a Dash 5 delivered in February 2007 – enjoyed lots of attention. However, the departure of this first machine, just three years later after 12,000 hours, went virtually unnoticed. Relieved from its prime-moving role, the excavator was quietly shipped out of the country to South America.

Komatsu PC2000

Recently Celtic Energy replaced the first Dash 5 with another Japanese-made digger – which is the very first Komatsu PC2000-8 in Britain. (See Chapter 5 for technical information on this machine.) The 200 tonne Dash 8 earns its keep at Nant Helen, one of three Celtic Energy surface coal mines, which lies on the southern edge of the Brecon Beacon National Park.

It is an extensive site that has excavated a staggering 60 million cu m of overburden and taken out over 3 million tonnes of coal since surface mining began more than a decade ago. With two years left to run, another 8 million cu m of overburden needs removing to reach the

The very first Komatsu PC2000 in Britain is proving a valuable asset at Nant Helen.

In the right conditions the PC2000 extracts as much as 600 to 700 cu m an hour.

last 800,000 tonnes of coal. The best seam is almost 2.5 m thick and lies at a depth of 110 m.

The seams run up and down all over the place making the coal very difficult to reach. The mining process is further complicated by a band of extremely tough rock which is the job of a pair of RH120Es, with 15 cu m buckets, and an older RH120C. Some of the more confined areas are tackled by a Komatsu PC1250.

Despite its smaller 12 cu m bucket, the PC2000 is still a serious digger capable of filling a Cat 777 in five attempts – four on a good dig. The nimble excavator is ideally suited for tackling areas that are difficult to reach.

'The PC2000 is built well and offers great breakout forces,' reckons Alan Roberts, the general foreman at Nant Helen. 'It is an ideal size for this site.'

The PC2000 is also surprisingly quiet even when standing just a short distance away. The quiet-running engine is a big bonus since the site is not far from a local community.

After spending many hours behind the controls of the EX1900-5, it did not take operator Norman Harper long to get to grips with the newcomer. After 1,000 hours in the seat, he confesses he loves the sturdy build of the

Komatsu excavator. 'It is a very stable machine and I think that it digs as well as an RH120,' he says.

So far there are smiles all round; everyone is pleased with the performance of the excavator and with the service and back-up provided by Komatsu. Apart from a minor problem with the greasing system – which was quickly resolved – there have been no other issues.

Unlike the larger PC3000, which is handled by KMG Warrington, the PC2000 is looked after by Marubeni-Komatsu UK. Although it may seem a bit of an odd arrangement, Celtic Energy group director Huw Richards confirms this company is pulling out all the stops and making a real effort to prove the new excavator.

While genuinely impressed by what he has seen so far, Huw says he will not accept any drop in standards during the 12,000-hour warranty period.

'If Marubeni-Komatsu get it right, then we will tell everyone. However, it works the same way if they get it wrong.'

Although Celtic Energy has changed brand, they continue down the 200-tonne excavator route. Huw is surprised that other British surface mines have not latched on to the cost-per-tonne savings provided by

This RH120C, used as a spare machine, is still going strong at Nant Helen after 24,000 hours.

The close proximity to the Brecon Beacons National Park means that it is crucial for the mine to keep on top of dust.

Really tough digging is reserved for a pair of RH120Es. At full capacity, each one is able to move as much as 1,000 cu m an hour.

The Komatsu PC1250 is a great excavator to scrape off the final layer overburden to expose the coal.

This RH120E is currently out of action for a spot of maintenance. It carries a 15 cu m bucket.

this class of excavator when loading a hundred-tonne truck.

'It is a good mix,' he says. 'A 300-tonne excavator is approximately 30 per cent more expensive to purchase and operate than a 200-tonner, but in our tough geological strata we do not always get 30 per cent more production. A 200 tonne excavator is the perfect tool for these conditions.'

With a large chunk of Welsh land in its portfolio, Celtic Energy is cautiously optimistic of the future and it is in the process of obtaining permission to extend Nant Helen in order to remove another 1.8 million tonnes of coal.

This optimism follows the fact that Selar – another of Celtic Energy's three big surface mines – has been awarded the green light to extend. (See Chapter 3.) During the next five years the company expects to dig up a million tonnes of coal at Selar.

Now the attention has turned to Nant Helen and the

future of 127 fulltime employees, 80 per cent of whom live within a 5-mile radius of the site. In an area where unemployment is the rule rather than exception, the prospect of unnecessary job loss can be used as a key bargaining tool by the surface mining operation. Celtic Energy is a vital employer in the area and they have a good relationship with the local council. The three Celtic Energy surface mines and washing plants provide 300 direct jobs and another 150 ancillary jobs down their supply chain.

'If we do not get the green light to forge ahead at Nant Helen, then up to 150 jobs are at stake,' he warns. 'It will be a bit like closing a factory.'

Hitachi EX1900

The Welsh coal mining company also hopes to get permission to extend at East Pit East Revised. At the northern edge of the coal belt, the coal is shiny and clean and if you rub small lumps of it together in your

A Hitachi excavator loads coal. Nant Helen currently extracts around 1,800 tonnes of a day.

hands, it sounds like marbles. Drop it on the ground and it does not break. The coal on this edge of East Pit East Revised is described as the best quality coal in Wales. It costs around £150 a tonne!

Some of Celtic Energy's coal travels to Tata Steel at Port Talbot as well as Aberthaw power station along what used to be the Brecon–Neath passenger rail link.

East Pit East Revised is just visible from Nant Helen across the far side of the valley. Much has changed at this site in the last year. The huge lake at the bottom of the void is no longer as deep because a number of large pumps spent a year draining the water at a rate of 900 cu m per hour.

With the water level lowered, a team of large earthmovers are getting on with the task of shifting millions of tonnes of overburden. Moving in a westerly direction, the big machines are digging down through the first coal seams, some of which are just 150 mm thick. While there is not much coal in the top three seams, there is too much to throw away. With overburden ratios often above 20:1, every tonne of coal helps to cover astronomical diesel prices as they dig down to the more valuable seams.

In the current phase of East Pit East Revised, coal extraction can be as low as 2,500 to 3,000 tonnes a

week, increasing to 10,000 to 12,000 tonnes when they tap into thicker seams.

Almost as if they are saving the best for last, the 2.5 m base seam could take nearly two years to access as it lies at a depth of 160 m – which is 30 m below the current water line.

East Pit East Revised is home to a Komatsu PC3000-1 and a pair of Komatsu PC1250-8s as well as the older Dash 6 version of the Hitachi EX1900 – the sole survivor of its kind in Britain.

Commissioned in September 2008, the excavator continues to do exactly what it was bought to do, dig and load large volumes of material. However, now nudging 10,000 hours and nearly out of warranty, this excavator is soon to be replaced with a second PC2000. Now that the machine is up for sale, it is quite possible that the Dash 6 may already have left the country by the time this book is published.

Hitachi enjoys a measure of success with large mining excavators in Finland and Spain. Sadly, the toehold it has enjoyed with the two EX1900s for the past few years in South Wales appears to have been short-lived.

The burning question is why were the first pair of EX1900s in Britain relieved from their prime-moving role so soon?

It is a difficult and politically sensitive question to answer. Huw confirms that price is important when concluding a deal and the favourable exchange rates at the time of purchase meant they got a good deal on both Hitachi excavators.

However, service and back-up are equally crucial. 'We need people and we need support,' he says. He feels they have been somewhat let down on both counts.

There is of course a commercial reason why the two EX1900s were axed so soon after the 10,000-hour threshold and this is due to warranty. Up until the end of warranty, the maintenance costs are low for the Hitachis. It would seem that Celtic Energy is not keen to pick up increases in running costs.

The future of the last surviving Hitachi EX1900 in Britain hangs in the balance. It is for sale and will probably be exported.

Three-tiered Approach

Historically, Celtic Energy has been a big fan of the Demag H 185, its successor the Komatsu PC3000-1, and the Terex O&K RH120. Now, though, the Welsh mining group appears to have selected a single brand for the future and one that could leave the RH120 out in the cold.

Huw Richards thinks that Komatsu's present-day PC3000 has caught up with the RH120. He is of the opinion that the RH120 may be starting to lag behind other manufacturers' models in technological advance.

Compelling evidence in favour of Komatsu replacements can be found not far from the workshop at East Pit East Revised where the second of three brand-new PC3000-6s is nearing completion. The three newcomers – two backhoes and a face shovel – were bought to relieve a couple of older PC3000-1s from their prime-moving roles. The first backhoe excavator recently started work at Selar following the successful outcome of the application for extension.

Celtic Energy's preference for Komatsu is possibly fuelled by the presence of Miller Argent's four PC3000-6s, located not far away at Merthyr Tydfil. The decision-making process was also no doubt aided by the thousands of hours of reliable service previously provided by their older PC3000-1s.

Whatever the reason, it will not be long before two of Celtic Energy's three big coal mines each have one PC3000, one PC2000 and one PC1250. This three-tiered approach gives them a good spread of digging options. In the right conditions, the PC3000 can shift 1,000 cu m an hour, just what was needed for mass overburden, while the PC2000 provides the mines with a good multi-purpose muck-shifter, averaging 600 to 700 cu m an hour. The PC1250 is viewed as a utility machine, ideal for scraping the last metre of overburden from a seam without forcing it into the coal.

The three-tiered approach sounds good on paper but it remains to be seen how it works in practice. Will Celtic Energy get the reliability and service that it is looking

for? If the approach is successful, then other British surface mines may take a closer look at a similar package and at running a 200-tonne excavator in particular.

Green Light for Cat

During the course of this book I have made numerous references to Caterpillar and its bid for Bucyrus International, an acquisition that was pending approval from the regulatory bodies. And, sure enough, just as I was closing this Postscript along came the long-awaited news that the multi-billion dollar bid had finally been approved.

The acquisition is huge news because it provides Cat with an unrivalled global range of mining machinery, both above and below ground, with very little duplication. Up to now Cat has played a minor role in the underground mining business. Suddenly, due to the purchase of Bucyrus, it is about to take a huge step forward in this rapidly growing sector. Similarly, the company will take a giant leap in the surface mining sector with an unbelievably large range of products that even includes blast-hole drilling rigs.

Prior to the acquisition, if you look purely at global construction machine sales, it is obvious that Komatsu has worked hard to become regarded as being on a par with Caterpillar. However, the approval of this latest mega-deal means Cat is now in a league of its own. If Komatsu wants to keep pace, it will need to react quickly.

In the meantime, Caterpillar is moving swiftly to add its newly acquired machinery to its equally new Mining Equipment Group. All Bucyrus machines have already been given a provisional Cat model number and the promise of Cat colours seems to be imminent.

The only models that are the exception to the rule are the Bucyrus AC drive mining trucks. The five-model range offers payloads ranging from 136 to 363 tonnes and will instead be rebranded as Unit Rig. At face value, these seem to be the only products upsetting the synergy of this mega-marriage. As Cat already offers a hugely successful range of mining trucks there would appear to be a bit of overlap but this is incorrect. Cat makes mechanically driven versions, while Bucyrus has specialised in the AC electric route.

During the past two years Caterpillar has invested a fortune in the research and development of large electrically driven trucks. The first of these to hit the market is the giant 313 tonne capacity 795F AC. (See Chapter 13 on Sweden.) Still, it remains to be seen what plans Cat has for AC technology in the future.

Incorporating the Bucyrus International product line is a massive operation that presents Cat with major challenges. At the time of writing the details were sketchy, but initially they plan to operate with a dual distribution. Bucyrus will continue to produce machines with direct support from current Bucyrus employees, while Cat products will continue to come from Cat dealers, with support from its Global Mining department.

The concept rendering of a Cat 7495 electric rope shovel, formerly a Bucyrus 495HR2, loading a 797F mining truck.

However, over the course of the next year, it is anticipated that Cat dealers will have an increasingly important role to play with all product lines from sales to service. The transition will occur in phases based on the mining business opportunity and the population base of current Bucyrus products.

Exciting as the deal is, we are unlikely to see many of Caterpillar's newly acquired machines – such as the trio of massive 8000 series walking draglines – in Europe. Similarly, it remains to be seen whether we will ever see any of the seven-strong fleet of Bucyrus electric rope shovels in Europe. The flagship 7495 electric rope shovel, which is part of the re-labelled Cat 7000 series, can carry a monstrous 61.2 cu m dipper that easily lifts well over 100 tonnes at a time.

However, a couple of machines, such as the RH120E excavator, are of great interest to the UK. Potential customers are still getting used to the change of ownership and colour scheme from Terex to Bucyrus. Now it is just a matter of time before this popular earthmover changes colour yet again.

The exterior livery will not be the only change since, as discussed above, the new provisional model numbers are already out. Cat stresses that these numbers could change but for now the RH120E/RH120E FS (Front Shovel) is called the 6030/6030 FS, and is officially part of the Caterpillar 6000 series.

The new range encompasses all former Bucyrus RH excavators, right up to the 980-tonne RH400, which has been renamed the Cat 6090. Power to this phenomenal digger still comes from a 4,500 hp engine, which carries a bucket capable of lifting over 90 tonnes of material at a time!

Lower down the weight scale there are a couple excavators that could possibly provide Cat with additional business opportunities; these are the often overlooked 105 tonne RH40 and 172 tonne RH90. Provisionally the two excavators will be known as the Cat 6015 and the Cat 6018, but whether Cat can boost the popularity of these two excavators remains to be seen.

New Contenders

Monstrous earthmovers capable of lifting as much as 100 tonnes at a time are not yet in Hyundai's product list but this company is slowly moving up the excavator weight scale. Just a couple of years after it launched its very first 80 tonne excavator comes news that it is currently trialling a number of even larger 120 tonne machines.

Two of the initial trio of machines, dubbed the Hyundai R1200-9s, are believed to be working in New Zealand, while the third is thought to be undergoing tests at a Russian coal mine. While the technical specification remains largely under wraps, power to these machines is providing by a Cummins QSK23 engine capable of 741 hp and rumour has it that the R1200-9s carry a standard bucket of around 7 cu m.

Hyundai is not the only new contender looking to go further up the weight scale because Doosan, a South Korean company that has made huge progress in Europe during the past five years, recently launched its first 70 tonne excavator. It may not be too long before the well-established names of Cat, Liebherr, Hitachi and Komatsu are not the only choices when it comes to sourcing a new heavyweight excavator.

The important question here is: how far do these new players intend to go up the weight scale?

Perlini in the UK

Finally, on the subject of new contenders, this book began with a peek at Perlini, the Italian rigid dump truck maker, so it is fitting that this Postscript finishes with another reference to the brand.

The first new Perlini quarry trucks to be sold in the UK for many years were recently put to work, following Volvo's appointment as an official dealer. While the pair of DP 405s are certainly not the largest in the range, they are definitely a good starting point. Who knows, this purchase may ultimately pave the way for even larger Perlini trucks in the future.

De Rijp, The Netherlands
July 2011

This studio generated image provides a hint of what the Bucyrus 8750 dragline may look like as a Cat 8750.

This concept rendering of a Bucyrus RH200 hydraulic excavator reveals what it may look like as a Cat 6050.

1 Italy

Cat 385
Cat 5080B
Hitachi EX1100-3
Hitachi EX1200
Hitachi EX1200-5D
Hitachi EX1200-6
Liebherr R 984C
Perlini DP 405
Perlini DP 705
Perlini DP 905
Perlini DPT 70

2 The Netherlands

Blockbuster
Cat 345
Cat 385 (Condor)
Cat 385 Long Reach
Cat 773B
Cat 775B
Cat 988F
Hitachi EX1200
Komatsu HD605
Komatsu PC3000
Liebherr R 984
Liebherr R 994
Liebherr R 994B
Liebherr R 9250
Liebherr R 9350
Ruston-Bucyrus 150-RB

3 Wales

Cat 385
Cat 777F
Demag H 185
Demag H 185S
Demag H 255S
Hitachi EX1900-5
Hitachi EX1900-6

Komatsu PC1250-8
Komatsu PC2000
Komatsu PC3000
Komatsu PC3000-1
Komatsu PC3000-6
Water Cannons

4 Finland

Cat 375
Cat 385C
Cat 777F
Cat 988F
Cat 992G
Dresser Haulpak 4000
Euclid R190
Hitachi EH1100
Hitachi EH1700
Hitachi EH3500ACII
Hitachi EH3500DC
Hitachi EX800H
Hitachi EX1100
Hitachi EX1200-5
Hitachi EX1900-5
Hitachi EX1900-6
Hitachi EX3600-6
Kobelco SK1340
Komatsu 730E
Komatsu HD785
Komatsu PC4000
Komatsu WA700
Komatsu WA800
Komatsu WA900
Komatsu WA1200-3
Sleipner Transport System

5 Spain

Cat 785C
Cat 789
Cat 994

Cat 5080
Cat 5090
Cat 5110
Cat 5130B
Cat 5130ME
Cat 5230
Cat D10
Demag H 95
Demag H 185
Demag H 185S
Demag H 255S
Hitachi EH1100
Hitachi EX3500-5
Hitachi EX3600-6
Hitachi EX5500-5
Komatsu D475A
Komatsu D575A
Komatsu HD785
Komatsu HD1200M
Komatsu PC1100
Komatsu PC1100-6
Komatsu PC1250-7
Komatsu PC1500-1
Komatsu PC1600
Komatsu PC1800-6
Komatsu PC2000-8
Komatsu PC3000
Komatsu PC4000
Liebherr R 984
Liebherr R 994
Liebherr R 994B
Liebherr R 9350
Liebherr R 995
Liebherr T 252
Liebherr T 282C
Liebherr TI 274
Perlini DP 705
Terex O&K RH120E
Terex O&K RH200

6 England

Bucryrus-Erie 1150-B (Oddball)
Bucyrus-Erie 1260-W (Chevington Collier)
Cat 385
Cat 777
Cat 785B
Cat 789
Cat 992G
Cat 5090B
Cat 5130B
Cat D9
Cat D10R
Cat D11R
Demag H 111
Demag H 185
Demag H 485
Hitachi EX3500
Komatsu HD605
Komatsu HD785
Komatsu WA800
Komatsu WA900
Marion 7800
O&K RH40
O&K RH60
O&K RH75
O&K RH75C
O&K RH120
O&K RH170
O&K RH200
O&K RH300
Terex O&K RH120E
Terex O&K RH200
P&H 1200
P&H 757 (Ace of Spades)
Poclain 1000CK
Poclain EC1000
Ransomes & Rapier W1400 (Sundew)
Ransomes & Rapier W2000
Ruston Bucyrus 195-RB
Terex TR100

7 Luxembourg

Cat D10T
Cat D11N
Cat D11R CarryDozer (CD)
Cat D11T

8 Switzerland

Cat 771C
Cat 771D
Cat 775F
Cat 990
Cat 990H
Cat 992C
Cat 992K
Cat D8T
Cat D10
Cat D11T

9 Scotland

Bucyrus RH120E
Bucyrus-Erie 1260
Bucyrus-Erie 195-B
Bucyrus-Erie 380-W
Cat 16M
Cat 777
Cat 785D
Cat 834H
Cat 992C
Cat D9
Cat D10R/T
Demag H 135S
Demag H 185
Demag H 185S
Demag H 255S
Demag H 285S
Demag H 485
Demag H 655S
Komatsu PC1250
Komatsu PC1400
Komatsu PC2000
Komatsu PC3000-6
Komatsu PC4000
Komatsu PC8000

Liebherr R 994
Liebherr R 9250
Liebherr R 9350
O&K RH40
O&K RH120C/E
Terex RH120E
Terex TR100

10 France

Bucyrus-Erie 295
Cat 631
Cat 657G
Cat 657E
Cat 777
Cat 854K
Cat 988H
Cat 990H
Cat 992G
Cat 993K
Cat 994F
Cat 5130
Cat D9R
Cat D10
Cat D11
Komatsu HD985
Komatsu WA600
Komatsu WA900
Liebherr RT 1000
Liebherr R 984
Liebherr R 991
Liebherr R 991HD
Liebherr R 992
Liebherr R 994
Liebherr R 994B
Liebherr R 995
Liebherr R 996B
Liebherr R 9100
Liebherr R 9250
Liebherr R 9350
Liebherr R 9400
Liebherr R 9800
Lima 2400B
Marion 7400

O&K RH120C
Poclain 61CK
Poclain 90CK
Poclain 350CK
Poclain 400CK
Poclain 600
Poclain 600CK
Poclain 610CK
Poclain 1000CK
Poclain 1000CK M1
Poclain 1000CK M2
Poclain EC 1000

11 Norway
Broyt X52
Cat 789B
Cat 789C
Cat 966E
Cat 988H
Cat 992
Cat 992G
Hitachi EX1800
Komatsu HD785-6
Komatsu HD1200
Komatsu PC600
Komatsu WA800
Komatsu WA900
Komatsu WA1200
LeTourneau L-1000
LeTourneau L-1100
O&K RH120C
P&H 1900
Terex O&K RH170
Unit Rig MT-3600

12 Belgium
Cat 992G
Cat 994
Cat 994F
Demag H 255S
Dresser Haulpak 445E
Dresser Haulpak 4000

Komatsu HD785
LeTourneau L-70
LeTourneau L-700 LeTro Loader
LeTourneau L-1000
LeTourneau L-1100
LeTourneau L-1200
LeTourneau L-1400
LeTourneau L-1800
LeTourneau L-2350
LeTourneau SL-40
Marathon LeTourneau L-800
Terex O&K RH120E
Unit Rig Mark 36
Unit Rig MT-2700
Unit Rig MT-2700
Unit Rig MT-3600

13 Sweden
(Unit Rig) Lectra Haul MT-4000
Bucyrus 495BII
Bucyrus-Erie 295-B
Cat 24M
Cat 789 (Water Truck)
Cat 793 (B, C, D)
Cat 795F AC
Cat 994
Cat 994F
Demag H 485
Komatsu PC5500
Komatsu PC8000
Komatsu WA1200
LeTourneau L-1350
LeTourneau L-1800
Marion 7400
P&H 4100
P&H 4100A
P&H 4100C
Ruston-Bucyrus 150-RB
Volvo EC700
Volvo L330E

14 Germany
Benrather Baggerfabrik B 300
Benrather Baggerfabrik B 400
Benrather Baggerfabrik B 504
Bucket Wheel Excavator 285
Bucket Wheel Excavator 288
Bucyrus RH200
Bucyrus RH400
Cat D8
Cat D9
Clark-Michigan 475
Demag E30
Demag H 65
Demag H 101
Demag H 185
Demag H 185S
Demag H 241
Demag H 285S
Demag H 455S
Demag H 485
Faun Rigid Dump Truck
Komatsu PC3000
Komatsu PC4000
Komatsu PC5500
Komatsu PC8000
Komatsu WA800
O&K RH5
O&K RH90
O&K RH120
O&K RH170
Poclain CK300

About the Author

Steven Vale is the European correspondent for *Earthmovers* magazine. He lives with his wife Hilda and two children, Carmina and Cameron, in an old fishing village called De Rijp, a short drive north of Amsterdam in the Netherlands.

He enjoys travelling the length and breadth of Europe documenting the biggest earthmoving machines. However, now that his first book is finished, Steven hopes to spend more time with his parents, John and Heather, who live in Northamptonshire. The rest will certainly be well deserved because Steven also invests large portions of his time in helping to make some of the 'Massive Machines' series of DVDs for Old Pond Publishing.

With more titles in the pipeline and limitless enthusiasm for the subject, Steven plans to continue the *Walking with Giants* theme with a book on the largest and tallest demolition machines. There is still plenty to keep him busy in Europe.

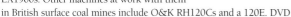

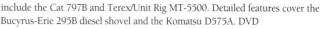